2006 IEEE WORKSHOP ON MICROELECTRONICS AND ELECTRON DEVICES

Boise Centre on the Grove
April 14, 2006

WORKSHOP SPONSORS
IEEE Electron Devices Society – Boise Chapter
Boise State University, College of Engineering
Electrochemical Society (ECS)
IEEE EDS Region 6 / SRC
IEEE Boise Section
Micron Foundation

PARTICIPATING INSTITUTIONS
American Semiconductor
AMI Semiconductor
Boise State University
Micron Technology
Montana State University/ Auburn University
Oregon State University
Sandia Laboratories
SEMATECH
TU Berlin
University of Idaho
University of Utah
University of Tehran
Washington State University

2006 IEEE Workshop on Microelectronics and Electron Devices

Copyright © 2006 by the Institute of Electrical and Electronics Engineers, Inc.
All rights reserved.

Copyright and Reprint Permission

Abstracting is permitted with credit to the source. Libraries are permitted to photocopy beyond the limit of U.S. copyright law, for private use of patrons, those articles in this volume that carry a code at the bottom of the first page, provided that the per-copy fee indicated in the code is paid through the Copyright Clearance Center, 222 Rosewood Drive, Danvers, MA 01923.

Other copying, reprint, or reproduction requests should be addressed to:
IEEE Copyrights Manager, IEEE Service Center, 445 Hoes Lane, P.O. Box 1331, Piscataway, NJ 08855-1331.

IEEE Catalog Number:	06EX1374 (Softbound)
	06EX1374C (CD-ROM)
ISBN Softbound:	1-4244-0373-1

Additional copies of this publication are available from

IEEE Operations Center
P.O. Box 1331
445 Hoes Lane
Piscataway, NJ 08855-1331 USA

1-800-678-IEEE
1-732-981-1393
1-732-981-9667 (FAX)
email: customer.services@ieee.org

2006 IEEE Workshop on Microelectronics and Electron Devices

Boise, Idaho
14 April 2006

IEEE Catalog Number: CFP06564-POD
ISBN: 978-1-42440-373-8

WMED 2006 – Table of Contents

Preface vii

Program ix

Session Circuit Design

S1p1   Indirect Feedback Compensation of CMOS Op-Amps 3
Vishal Saxena, R. Jacob Baker

S1p2   A CMOS Low Voltage Down-Converter Mixer for Sub 1V Applications 5
M. B. Vahidfar, O. Shoaei

S1p3   Shared Multiplier Design of a Digital Filter on a High-Temperature
FPGA Module 7
Bijan Houle, Vishu Gupta, Kevin Buck, Herbert L. Hess, Gregory Gregory, Randy Normann

S1p4   40 Gbps SiGe Pattern Generator IC with Variable Clock Skew
and Output Levels 9
Matthew J. Zahller, George S. La Rue

S1p5   High Speed Digital Input Buffer Circuits 11
Krishna Duvvada, Vishal Saxena, R. Jacob Baker

MEMS Devices and Sensors

S2p1   Polymer-based Thin Film Coils as a Power Module
of Wireless Neural Interfaces 15
S. Kim, K. Buschick, K. Zoschke, M. Klein, M. Toepper, D. Black, R. Harrison,
P. Tathireddy, F. Solzbacher

S2p2   Design and Fabrication of a MEMS Capacitive Chemical Sensor System 17
Vishal Saxena, Todd J. Plum, Jeff R. Jessing, R. Jacob Baker

S2p3   A High Sensitive Piezoresistive Sensor for Stress Measurements
in Packaged Semiconductor Die 19
Ahsan Mian, Jeffrey C. Suhling, Richard C. Jaeger

S2p4   Resistance Switching in SnxMnyTez-based Devices 21
Patrick K. Herring, Kristy A. Campbell

S2p5   Metal/Semiconductor Contacts for Organic Molecules 23
Divesh Kapoor, Justin B. Jackson, Mark S. Miller

Processing and Reliability

S3p1 Space Efficient ESD Methodology for Reliable High Volt Applications.............27
John J. Naughton, Matthew Tyler, Muhammad Anser

S3p2 Time-Dependent Dielectric Breakdown of a Recessed Channel
DRAM Access Device 29
T. Owens, D. Hwang, P. Vaidyanathan, K. Parekh

S3p3 Preliminary Study of NOR Digital Response to Single pMOSFET
Dielectric Degradation 31
T. L. Gorseth, D. Estrada, J. Kiepert, M. L. Ogas, B. J. Cheek, P.M. Price, R. J. Baker,
G. Bersuker, W. B. Knowlton

S3p4 Ultra-Low-Power Dynamic-Threshold Digital Circuits in the FlexFET
Independently-Double-Gated SOI CMOS Technology 33
S. Parke, K. DeGregorio, D. Hackler, D. Wilson

S3p5 Secondary Ion Mass Spectrometry Analysis of Wafer Contamination
Resulting from Gloved Hands 35
Wendy Morinville, Chantelle Krasinski

Poster Papers and Abstracts

S4p01 Thermal Noise Limits in Nanoscale Electronics...39
M. Mudrow, W. Wanalertlak, L. Forbes

S4p02 A Novel Triple Mode LNA Designed in CMOS 0.18 µm Technology
for Multi-standard Receivers 41
M. B. Vahidfar, O. Shoaei

S4p03 Power Dissipation and Temperature Variations in Nanoscale Devices...........43
W. Wanalertlak, M. Y. Louie, L. Forbes

S4p04 Micro-Sensor for Monitoring Oils 45
Brian M. Marx, Matthew Luke, Darryl P. Butt

S4p05 Integrated Silicon Nanowire Diodes 47
Justin B. Jackson, Sun-Gon Jun, Divesh Kapoor, Mark S. Miller

S4p06 Design of a MEMS Capacitive Chemical Sensor Based on Polymer Swelling.....49
T. J. Plum, V. Saxena, J. R. Jessing

S4p07 Introduction to Modeling an Imaging System with Human Detection
of Artifacts 51
Vinesh Sukumar, Doug Warner, Patrick Doherty, Herbert Hess, Ken Noren, Steve Krone

S4p08 Comparative Study of TaN-TiN and TiN Gate Stacks
for Thermally Stable PFETs 53
Nirmal Ramaswamy, Allen Mcteer, Venkat Ananthan, Nanda Palaniappan,
Tim Owens, Sanh Tang, Ravi Iyer, Shixin Wang, Chandra Mouli

S4p09 Dependence of Si3N4 Film Properties on Precursor Chemistry 55
Fernando Gonzalez, Shyam Surthi, Parag Banerjee

S4p10 Photo Sensitivities in a 0.35 µm 18 V PDMOS Technology............................57
Brett Williams, Mike Thomason, Chuck Belisle

S4p11   Silicon Solar Cells Using Backside Contacts
with Through-Wafer Interconnects 59
Aaron Erbe, A. J. Moll

S4p12   CMOS Imager Pixel Design for Space Applications 61
Mark Elgin, Dede Russell, Matt Katula, Ryan Paulsen, Stephen Parke

S4p13   Characterization of Negative Differential Resistance
in Chalcogenide Devices Containing Silver 62
Armand Bregaj, Kristy A. Campbell

Author Index 63

Preface

The Boise Chapter of the IEEE Electron Devices Society welcomes you to this fourth annual Workshop on Microelectronics and Electron Devices (2006 WMED), Friday, April 14, 2006 at the Boise Centre on the Grove. This workshop is co-sponsored by IEEE Boise Section, IEEE EDS Region 6, Boise State University, Electrochemistry Society, and the Micron Foundation.

The previous three workshops have been very successful, attracting students, faculty, and industry researchers from throughout the Northwest region for a day of engaging tutorials, invited speakers, posters, technical presentations, and professional networking. We hope you will make it an annual tradition to attend and participate in this Workshop.

The challenge of a regional workshop is whether to make the focus diffuse and of interest to a broad set of participants, or rather to focus on an apropos topic or issue. Given the state of the industry and the demographics of our constituency, the organizers have chosen the latter. Specifically, for this year, the WMED has more topics of interest to memory technologists, although this is certainly not universal. Future Workshops will focus on other topics.

We are honored to have Al Fazio give this year's keynote talk on the future on non-volatile memory. The market growth and technical challenges facing the non-volatile memory industry, coupled with Al's many years of development experience, makes this a particularly exciting topic.

Keynote
- "Future Directions of Non-Volatile Memory Technologies"
  Al Fazio; Director, Memory Technology Development, Intel Corporation

Given the focus of this year's Workshop, we are pleased to have two very experienced industry executives give tutorials.

Tutorials
- "Current Reliability and Failure Analysis Issues for High Volume Memory Products"
  Christopher L. Henderson; President, Semitracks, Inc.

- "3D Memory Packaging Technologies & Trends"
  Nozad Karim; VP, App. Engrg. & Characterization, Amkor Technology

In addition, we have three excellent invited speakers with talks ranging from process engineering to device applications.

Invited Talks
- "Preserving Low-k Dielectrics During Photoresist and Etch Residue Removal"
  Dennis Hess; William W. LaRoche, Jr. Chair, School of Chemical & Biomolecular Engineering, Georgia Tech

- "Lest We Never Forget: A Historical Perspective of Space Data Recorders; A Prediction for the Future"
  Karl Strauss; Senior Engineer, Flight Avionics Systems, Jet Propulsion Laboratory

- "Physics Based Modeling and Design of Chips, Devices, and Nanostructures: Present and Future"
  Neil Goldsman; Professor, Department of Electrical Engineering; Univ. Of Maryland at College Park

Following the keynote session are three parallel Technical Sessions with five contributed papers in each:

- Advanced Processes & Reliability
- Microelectronic Devices & Circuit Design
- Microelectromechanical Systems (MEMS)

Finally, the Poster Session contains several excellent posters with their authors available for questions and answers.

This year's organizing committee is as follows:

General Chair:	Roy Meade, Micron Technology, Inc.
Technical Program Chair:	David Hwang, Micron Technology, Inc.
Publications Chair:	Kris Campbell, Boise State University
Registration Chair:	Steven Groothuis, Micron Technology, Inc.
Publicity Chair:	Fernando Gonzalez, Micron Technology, Inc.
Local Arrangements:	Gail Hawkins, Micron Technology, Inc.
	Virginia McGraw, Micron Technology, Inc.
Treasurer:	Zhiping Yin, Micron Technology, Inc.

The committee would like to thank the following professors for supporting the workshop at their respective institutions: Jake Baker, and Steve Parke at Boise State University; Herb Hess at the University of Idaho; Leonard Forbes at Oregon State University; Florian Solzbacher at the University of Utah; George LaRue at Washingtion State University. The committee would also like to thank the R&D managers at Micron Technology, Inc., and AMI Semiconductor for their support and participation in this Workshop.

Finally, we want to thank all those who have helped make this year's WMED a success. In particular, the financial support of the Micron Foundation and the ECE Department of Boise State University is sincerely appreciated.

Roy Meade
WMED 2006 General Chair

WMED Program – April 14, 2006

Tutorials

Time	(Firs/Pines Room)	(Willows Room)
9:15-11:00	**Current Reliability and Failure Analysis Issues for High Volume Memory Products** — Christopher L. Henderson; President, Semitracks, Inc.	**3D Memory Packaging Technologies & Trends** — Nozad Karim; VP, App. Engrg. & Characterization, Amkor Technology
11:00-12:30	Lunch on your own in Downtown Boise	

Invited Speakers

Time	
12:30-12:45	**Welcome** *(Firs/Pines Room)*
12:45-1:30	*Keynote:* **Future Directions of Non-Volatile Memory Technologies** — Al Fazio; Director, Memory Technology Development, Intel Corporation
1:30-2:15	**Preserving Low-k Dielectrics During Photoresist and Etch Residue Removal** — Dennis Hess; William W. LaRoche, Jr. Chair, School of Chemical & Biomolecular Engineering, Georgia Tech
2:15-2:30	Break
2:30-3:15	**Lest We Never Forget: A Historical Perspective of Space Data Recorders; A Prediction for the Future** — Karl Strauss; Senior Engineer, Flight Avionics Systems, Jet Propulsion Laboratory
3:15-4:00	**Physics Based Modeling and Design of Chips, Devices, and Nanostructures: Present and Future** — Neil Goldsman; Professor, Department of Electrical Engineering, Univ. of Maryland at College Park
4:00-4:15	Break

Time	**Circuit Design** *(Willows Room)*	**MEMS, Devices, and Sensors** *(Cottonwoods Room)*	**Processing and Reliability** *(Firs/Pines Room)*
4:15-4:30	**Indirect Feedback Compensation of CMOS Op-Amps,** Saxena, *et al.,* BSU	**Polymer Based Thin Film Coils as a Power Module of Wireless Neural Interfaces,** S. Kim, *et al.,* U Utah/TU Berlin	**Space Efficient ESD Methodology for Reliable High Volt Applications,** Naughton, *et al.,* AMIS
4:30-4:45	**A CMOS Low Voltage Down-Converter Mixer for Sub 1V Applications,** Vahidfar, *et al.,* U Tehran	**Design and Fabrication of a MEMS Capacitive Chemical Sensor System,** Saxena, *et al.,* BSU	**Time-Dependent Dielectric Breakdown of a Recessed Channel DRAM Access Device,** Owens, *et al.,* Micron Technology
4:45-5:00	**Shared Multiplier Design of a Digital Filter on a High-Temperature FPGA Module,** Houle, *et al.,* U Idaho/Sandia	**A High Sensitive Piezoresistive Sensor for Stress Measurements in Packaged Semiconductor Die,** Mian, *et al.,* MT State/Auburn	**Preliminary Study of NOR Digital Response to Single pMOSFET Dielectric Degradation,** Gorseth, *et al.,* BSU/SEMATECH

Time			
5:00–5:15	Ultra-Low-Power Dynamic-Threshold Digital Circuits in the FlexFET Independently-Double-Gated SOI CMOS Technology, Parke, et al., BSU/American Semiconductor	Resistance Switching in $Sn_xMn_yTe_z$-Based Devices, Herring, et al., Micron Technology/BSU	40 Gbps SiGe Pattern Generator IC with Variable Clock Skew and Output Levels, Zahller, et al., WSU
5:15–5:30	Secondary Ion Mass Spectrometry Analysis of Wafer Contamination Resulting from Gloved Hands, Morinville, et al., Micron Technology	Metal/Semiconductor Contacts for Organic Molecules, Kapoor, et al., U Utah	High Speed Digital Input Buffer Circuits, Duvvada, et al., BSU
	Poster Session & Hors-d'Oeuvres *(Vestibule and Cottonwoods Room)*		
5:30–6:00	Micro-Sensor for Monitoring Oils, Marx, et al., BSU	Thermal Noise Limits in Nanoscale Electronics, Mudrow, et al., OSU	Comparative Study of TaN-TiN and TiN Gate Stacks for Thermally Stable PFETs, Ramaswamy, et al., Micron Technology
5:30–6:00	Integrated Silicon Nanowire Diodes, Jackson, et al., U Utah	A Novel Triple Mode LNA Designed in CMOS 0.18um Technology for Multi Standard Receivers, Vahidfar, et al., U Tehran/U Pavia	Dependence of Si_3N_4 Film Properties on Precursor Chemistry, Gonzalez, et al., Micron Technology
5:30–6:00	Design of a MEMS Capacitive Chemical Sensor Based on Polymer Swelling, Plum, et al., BSU	Power Dissipation and Temperature Variations in Nanoscale Devices, Wanalertlak, et al., OSU	Photo Sensitivities in a 0.35um 18V PDMOS Technology, Williams, et al., AMIS
5:30–6:00	Introduction to Modeling an Imaging System with Human Detection of Artifacts, Sukumar, et al., Micron Technology/U Idaho	CMOS Imager Pixel Design for Space Applications, Elgin, et al., BSU	Silicon Solar Cells Using Backside Contacts With Through-Wafer Interconnects, Erbe, et al., BSU
5:30–6:00		Characterization of Negative Differential Resistance in Chalcogenide Devices Containing Silver, Bregaj, et al., BSU	Preliminary Investigation of a High Dielectric Constant Material to Replace SiO_2 in Metal Oxide Semiconductor Devices to Improve MOSFET Reliability, Elgin, et al., BSU
6:00–6:30	**Awards & Conference Closing** *(Firs/Pines Room)*		

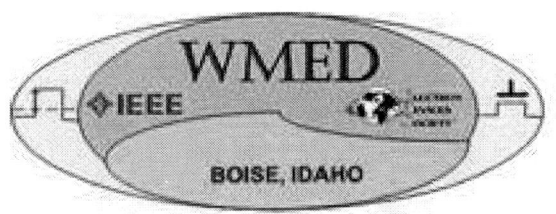

Session

Circuit Design

Indirect Feedback Compensation of CMOS Op-Amps

Vishal Saxena and R. Jacob Baker

ECE Dept., Boise State University, vishalsaxena@ieee.org

Abstract—**This paper presents the design of CMOS op-amps using indirect feedback compensation technique. The indirect feedback compensation results in much faster and low power op-amps, significant reduction in the layout size and better power supply noise rejection.**

Keywords- *CMOS, indirect feedback compensation, miller compensation, operational amplifier.*

I. INTRODUCTION

CMOS Op-amps are one of the important building blocks of modern integrated systems. The op-amps have been commonly stabilized using direct (or Miller) compensation in the past. This method achieves dominant pole compensation by pole-splitting due to Miller effect [1]. However, the connection of the compensation capacitance (C_c) between the outputs of the gain stages, leads to a right hand plane (RHP) zero. The RHP zero decreases the phase margin, and thus requires a larger C_c to compensate the op-amp. This in turn results in a decrease in the unity gain frequency ($f_{un} = g_{m1}/2\pi C_c$). Also the op-amp stability degrades when the load capacitance C_L becomes comparable to C_c as C_L must be much less than $g_{m2}C_c/g_{m1}$ for stability [2].

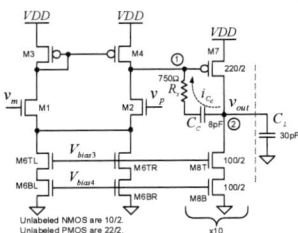

Figure 1. Two stage op-amp with miller (direct) compensation and zero-nulling resistor.

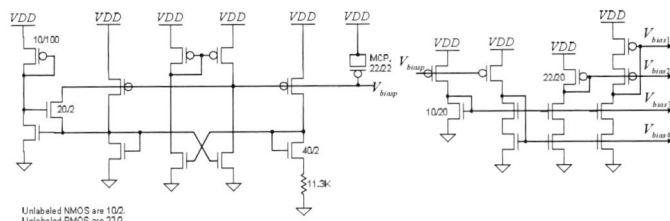

Figure 2. The self-biased reference ciruit used for baising the op-amps.

This paper introduces indirect feedback compensation technique which leads to much faster op-amps with significant reduction in the layout size. Fig. 1 shows a direct (Miller) compensated op-amp with an RHP zero-nulling resistor R_z. The op-amps presented in this paper are designed with AMI's CN5

(0.5μm) process, biased with a regulated drain BMR (Beta Multiplier Reference) bias circuit shown in fig.2 [3], and drive up to 30pF off-chip load.

II. INDIRECT FEEDBACK COMPENSATION

In a direct compensated two-stage op-amp, the current feedback through the compensation capacitor C_c can be approximated as $i_{C_c} \approx v_{out}/(1/j\omega C_c)$. By indirectly feeding this current to the output of the diff-amp, pole splitting and hence op-amp compensation can be achieved. Also by avoiding connecting the compensation capacitor directly to the output of the diff-amp, the right hand plane (RHP) zero is eliminated.

The compensation current can be fed indirectly to the output of the diff-amp using, 1) a common gate amplifier [2], 2) a cascode structure, and 3) MOSFETs laid out in series with one device operating in triode region [3].

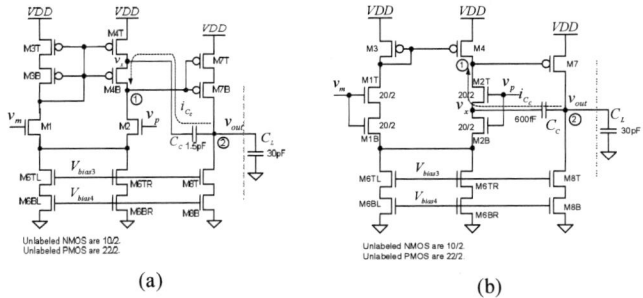

(a)   (b)

Figure 3. Two stage op-amp topologies with indirect compensation.

Fig. 3 shows op-amp topologies in which the feedback current is indirectly fed back to an internal low-impedance node. The low impedance node is created by laying out the MOSFETs in series, in which one of the devices is in triode. The topology in fig.3b results in a better PSRR (Power Supply Rejection Ratio) due to isolation of compensation capacitance from *VDD* and ground noise. As a guideline, the feedback current must always be fed back to a low-impedance internal node for high speed op-amps.

III. ANALYTICAL MODEL

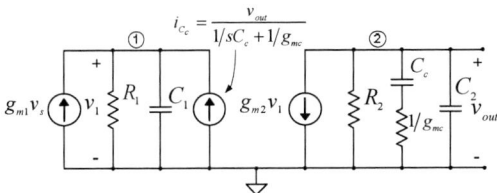

Figure 4. Model used to estimate bandwidth with indirect compensation.

Support of this project by EPA Contract No. X-97031101-0 is gratefully acknowledged.

To determine the frequency response of the op-amp with indirect feedback compensation, the generalized model seen in fig. 4 is employed.

Summing currents at node 1 gives,

$$-g_{m1}v_s + \frac{v_1}{R_1 \| \frac{1}{sC_1}} + \overbrace{\frac{v_{out}}{1/sC_1 + 1/g_{mc}}}^{i_{C_c}} = 0 \qquad (1)$$

where $1/g_{mc}$ is the resistance looking into the node v_x, where feedback current is injected.

For the output node (node 2), $v_{out} = -g_{m2}v_1 X_2$, where $X_2 = R_2 \| X_{C_L} \| (R_c + X_{C_c})$ is the total impedance on node 2, $R_c = 1/g_{mc}$ and $R_a = 1/g_{m1}$. $\qquad (2)$

On solving equations 1&2, the op-amp frequency response is estimated as,

$$\frac{v_{out}}{v_s} = \frac{-A_v\left(1 + \frac{jf}{f_z}\right)}{\left(1 + \frac{jf}{f_1}\right)\left(1 + \frac{jf}{f_2}\right)\left(1 + \frac{jf}{f_3}\right)}, \text{ where} \qquad (3)$$

$$f_z = \frac{g_{mc}}{2\pi.C_c}, \text{ which is a left hand plane (LHP) zero,} \qquad (4)$$

$$f_1 = \frac{1}{2\pi.g_{m2}R_1 R_2 C_c}, \qquad (5)$$

$$f_2 = \frac{g_{m2}R_c C_c}{2\pi.C_L(R_c C_c + R_1 C_1)} \approx \frac{g_{m2}C_c}{2\pi.C_L C_1}, \text{ and} \qquad (6)$$

$$f_3 = \frac{R_c C_c + R_1 C_1}{2\pi.R_1 C_1 R_c C_c} \approx \frac{1}{2\pi.R_c C_c}. \qquad (7)$$

The unity gain frequency (gain-bandwidth) of the op-amp is

$$f_{un} \approx f_1 A_v = \frac{g_{m1}R_1 g_{m2}R_2}{2\pi.g_{m2}R_1 R_2 C_c} = \frac{g_{m1}}{2\pi.C_c} \ (\approx f_z \text{ if } g_{m1} \approx g_{mc}). \qquad (8)$$

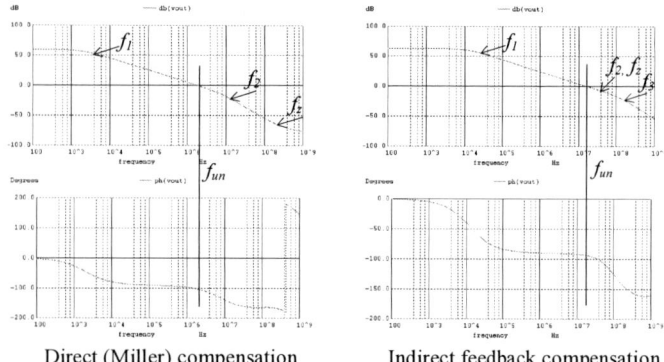

Direct (Miller) compensation   Indirect feedback compensation

Figure 5. Magnitude and phase responses of the opamps with direct (fig. 1) and indirect feedback (fig. 3b) compensation.

Fig. 5 shows the simulated frequency response for a direct (fig. 1) and an indirect feedback compensated op-amp (fig. 3b). The LHP zero (f_z) adds to the phase response and enhances the speed of the op-amp. Intuitively, at high speeds the phase shift through C_c causes the output signal to feed back and add to the signal at node 1. This positive feedback enhances the speed of the op-amp. The location of second pole (f_2) is at a considerably higher frequency. The net result is that, a higher value of unity gain frequency (f_{un}) can be set without affecting the stability of the op-amp. Moreover the load capacitance can be considerably large for a given phase or gain margin [3]. Thus the indirect feedback compensation results in much faster op-amp circuits and consumes significantly less layout area at the same power. The compensation capacitance value is reduced by 4 to 10 times, when indirect feedback compensation is used [4]. Also, the indirect feedback compensated op-amps are low power as the second stage need not be boosted much to push f_2 away from f_{un}.

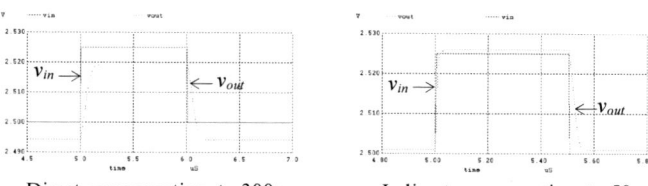

Direct compensation, t_s=300ns   Indirect compensation, t_s=50ns

Figure 6.   Step responses of an opamp with direct and indirect compensation.

Fig. 6 compares the step responses and settling times (t_s) for direct and indirect compensated op-amps and confirms that the latter is much faster than the former.

IV. PROGRESS AND FUTURE WORK

Two and three stage Op-amps with direct and indirect compensation are designed and being fabricated on a chip using AMI's CN5 process. The fabricated op-amps will be tested and the results will be compared with the analytical model.

V. CONCLUSION

The indirect feedback compensation is a practical and superior technique for compensation of op-amps and results in faster and low power op-amps with much smaller layout size. The indirect feedback compensation can also be extended to three (or more) stage op-amps [3].

REFERENCES

[1] P. R. Gray and R. G. Meyer, "MOS Operational Amplifier Design: A Tutorial Overview," *IEEE Journal of Solid-State Circuits*, vol. 17, pp. 969-982, Dec. 1982.

[2] B. K. Ahuja, "An Improved Frequency Compensation Technique for CMOS Operational Amplifiers," *IEEE Journal of Solid-State Circuits*, vol. 18, pp. 629-633, Dec. 1983.

[3] R. J. Baker, *CMOS: Circuit Design, Layout and Simulation, 2nd ed.* Boise, ID: Wiley-IEEE, 2005, pp. 531-538.

[4] R. J. Baker, "Design of High-Speed CMOS Op-Amps for Signal Processing," *IEEE/EDS WMED*, April, 2005.

A CMOS Low Voltage Down-Converter Mixer for Sub 1V Applications

M.B. Vahidfar[1], O. Shoaei[1]

[1]IC Design Center, ECE Department, University of Tehran
m.vahidfar@ece.ut.ac.ir

Abstract— **Scaling of CMOS technologies caused design of analog and especially RF blocks somehow challenging. The lower voltage supply is the most severe consequence of this reduction. In this paper a novel CMOS active mixer which can operate at low supply voltages by the use of switches connected to voltage supplies, is presented. Contrary to the conventional Gilbert mixer which is based on switching the RF current, in this mixer the load is switched on or off. The design is done in 0.18um CMOS technology for IEEE802.11b/g applications and the voltage supply can be scaled down to 0.8V which is compatible with 60nm CMOS technology. The simulation results show that the NF and IIP3 are better than 17 dB and 8 dBm respectively with 2.4GHz input RF signal. Moreover this design consumes lower power consumption in comparison to similar low voltage works.**

Key words: **Down converter CMOS mixer, low voltage, WLAN, linear transconductance.**

I. INTRODUCTION

Mixers are commonly used for frequency translation in radio frequency (RF) communication systems. A double-balanced Gilbert-type mixer (Figure 1) is the most mature mixer architecture widely used as the down converter in CMOS receivers, since it provides a high impedance input to the low-noise amplifier; yet is capable of driving a low impedance load at its output. However, due to high number of stacked transistors, this architecture can not be used in CMOS sub-microns technologies which work with supplies as low as 0.8 volt. A low noise, CMOS down converter mixer for IEEE802.11b/g applications which is compatible with 60nm CMOS technologies is discussed in this paper.

II. LOW VOLTAGE ARCHITECHTURES

The key problem of migration to sub-micron CMOS technologies comes from continual reduction in supply voltages, resulting in poor conducting switches in the mixer. Some recently published low voltage architectures [1,2,3], are introduced in this section and after highlighting their shortages, then proposed novel mixer is presented.

A. Folded architecture

The switched current Gilbert mixer can be enhanced for low voltage applications by using a folded architecture [1]. The

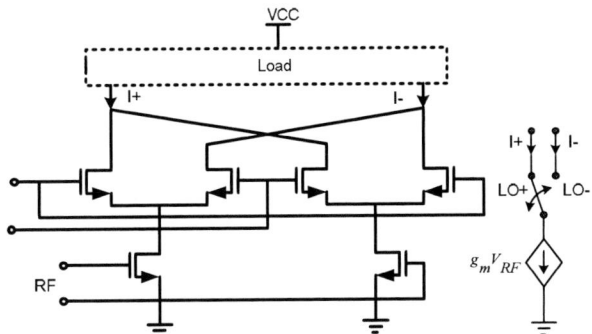

Fig. 1. Conventional Gilbert mixer based on switched current

main drawbacks of this mixer are as follows: 1) the peak-to-peak amplitude of LO signal is lower than supply voltages. 2) The circuit can not be easily biased, in spite of the fact that the linearity and especially IIP3 performance of this circuit is dependent on proper biasing of transconductor and switching section. 3) The switches are biased at non-zero drain current, which contributes more flicker noise to the output and increases the mixer NF.

B. Switched transconductor mixer

In this architecture the transconductor is switched on and off in each LO period [2]. The main drawback of this architecture is transition needed for the gm stage in each period to go from off-mode to its saturation mode and vice versa which degrades the linearity of mixer; therefore this architecture achieves poor IIP3.

C. Load switching mixer

Figure 2 shows the single balanced version of designed mixer enhanced for low voltage applications using load switching technique instead of RF current switching which is discussed in the next section.

III. DESIGNED LOW VOLTAGE MIXER

A. Operation

As it is shown in figure 2, the input RF signal of designed load switching mixer is applied to the transconductor. This signal is amplified at the output node when the switches connected to the output nodes are off. In the other words,

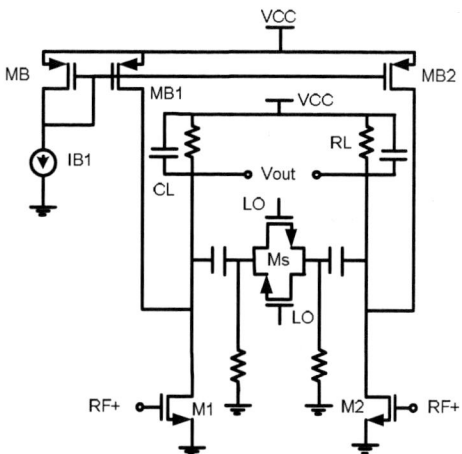

Fig. 2. Load switching mixer architecture

the circuit acts as an amplifier in this condition. When the switches are turned on, the output signal is killed. However due to switching pair non-idealities, the switching mechanism can be considered as a gain reduction mechanism which causes output signal experience lower gain in this condition. Shortly, the output signal has different gains in each half of clock period. Therefore the effect of the switching pairs on the output amplified RF signal can be modeled by a multiplication which causes the frequency translation of RF signal to the IF frequency band.

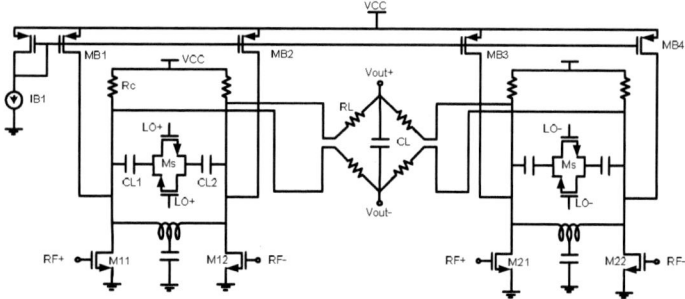

Fig. 3. Double balanced architecture of designed low voltage mixer

B. Implementation

To reduce the LO and RF signals injected to the output node, it is necessary to use a double balance architecture as shown in figure 3. This balanced circuit is completely symmetric and each side works in half of LO signal period. Each of the switches is implemented easily by a NMOS transistor. Due to asymmetry between gate-source and gate-drain capacitors of the NMOS transistor, the symmetry in the output nodes of mixer is degraded. To solve this problem, the source and drain voltages of each switch are fixed by a similar voltage, by biasing each switch in zero voltage. This approach leads to similar junction capacitance in source and drain nodes of switch, which are the main capacitance in the off mode. In the on mode, the switch directly go from off mode to triode region, therefore the capacitance of its gate with source and drain nodes (C_{gd}, C_{gs}) is equal to:

$$C_{gd} = C_{gs} = 0.5C_{ox} + C_{jb} + C_{ov}$$

Where, C_{ox}, C_{jb} and C_{ov} are MOS oxide capacitor, junction capacitor and overlap capacitor respectively.

The bias of mixer output nodes are defined by load resistors which are chosen for required conversion gain. Using resistors instead of active components is beneficial because it is flicker noise free, but they are voltage hungry. The circuit can work with the low voltage supply only if the bias current of the mixer is limited. On the other hand, the required mixer linearity can only be achieved by biasing the trans-conductance transistors in bias currents more than 1mA which is against using this architecture for low voltage applications. To solve this problem, a degree of freedom is added to the design by a redundant current path created paralleled with the output loads. The main portion of the bias current needed by transconductor is sourced by this current source. Referring to figure 2, this parallel path is made by MB, MB1-2 transistors.

To achieve a better NF performance, the switching transistors are biased in zero current and are decoupled from the output nodes by some capacitors (CL1-2). The flicker noise of the input RF section is also minimized by proper transistor sizing. The thermal noise of gm section is the main concern and has main contribution in output noise.

The mixer conversion gain is depended on input trans-conductance, LO amplitude and resistive load. The gain of the mixer can be reduced by adjusting both the bias voltage of input transistors and IB1 current simultaneously. The main advantage of this gain reduction mechanism, comparing with switching load resistors is saving output filter bandwidth in different gain modes.

Concerning the IIP3 performance, the pseudo architecture in the input transconductance section is better than LC degenerative architecture. The differential architecture is also not applicable due to consuming some voltage headroom.

IV. SIMULATION RESULTS

The simulation was done in a 0.18 um CMOS technology for 2.4GHz input RF signal and IEEE802.11b/g application. The mixer achieves 10dB conversion gain and 8dBm in-band IIP3 performance with 0.8 volt supply voltage. The NF of this designed mixer is 17dB with the total power consumption lower than 6.5mW. The mixer can also easily be used in CMOS sub-micron technologies such as 60nm with supply voltages as low as 0.8V.

REFERENCES

[1] Vojkan Vidojkovic, et all,, "A Low-Voltage Folded-Switching Mixer in 0.18-um CMOS" IEEE J. Solid-State Circuits, Vol. 40, NO. 6, June 2005.

[2] Eric A. M. Klumperink, et all , "A CMOS Switched Transconductor Mixer" IEEE J. Solid-State Circuits, Vol. 39, NO. 8, Augu 2004.

[3] Henrik Sjöland,, Ali Karimi-Sanjaani, and Asad A. Abidi, "A Merged CMOS LNA and Mixer for a WCDMA Receiver " IEEE J. Solid-State Circuits, Vol. 38, NO. 6, Augu 2003.

Shared Multiplier Design of a Digital Filter on a High-Temperature FPGA Module

Bijan Houle[1] Vishu Gupta[1] Kevin Buck[1] Herbert L. Hess[1] Gregory Donohoe[1]

houl6492@uidaho.edu gupt2491@uidaho.edu kmbuck@mrc.uidaho.edu hhess@uidaho.edu donohoe@ece.uidaho.edu

Randy Normann[2]

ranorma@sandia.gov

[1] *University of Idaho* [2] *Sandia National Laboratories*

Abstract— This paper outlines the design of a space-efficient digital filter for use in High-Temperature FPGA applications. It presents an implementation of a Butterworth filter design using VHDL: Shared Multiplier. It outlines the main signals and states used in the design. The Shared Multiplier design uses a single multiplier for each multiplication in the equation. The design is the smallest of the three. The results obtained by implementing the design on a FPGA are also presented. The Shared Multiplier approach is efficient in terms of demand of space on the FPGA.

I. INTRODUCTION

Sandia National Laboratories needed an extremely *space-efficient low frequency digital noise filter* for a circuit designed for geothermal data logging [1]. This circuit works at temperatures in excess of 250°C. To communicate with it, signals are sent over distances of up to a few kilometers, and testing has shown that low-frequency noise (≈1Hz) is a problem in the system. The filter will be placed *with existing designs* on a high-temperature FPGA alongside the circuit. The filter's requirements are as follows:

- IIR Filter
- Three poles
- Twelve-bit resolution[*]
- Sampling rate of 10Hz
- Bandwidth of 1Hz
- Multiplexing 6 signals

II. FILTER FRAMEWORK

We designed this filter using the FPGA [2] board and FPGA programmable software [3]. While this architecture is different from the High-Temperature FPGA architecture Sandia uses, our results still provide a useful impression of the relative size of the method.

Figure 1 shows the basic layout of the Shared Multiplier filter design. The filter has 4 input signals: Data_in, Start, Reset and Clock and two output signals: Data_out and Ready. Table I gives a brief description of these signals.

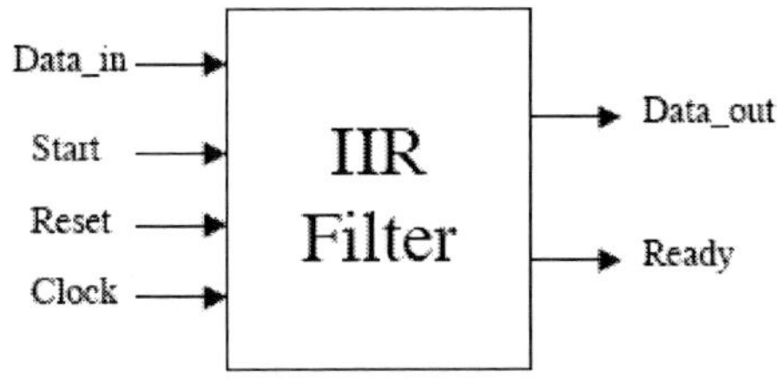

[*] The filter needs to interface to an existing 12-bit A/D converter.

Fig. 1: Layout of the filter.

TABLE I
SIGNAL DESCRIPTION

Data_in	12-bit	Input data
Start	1-bit	Start new filter sequence
Reset	1-bit	Synchronous Reset
Clock	1-bit	11 Mhz System Clock
Data_out	12-bit	Output Filtered Data*
Ready	1-bit	Ready/Busy indicator

We chose to use a Butterworth topology for our filter, in order to keep the design simple and small. The filter topology has 5 main states that are outlined below and shown in Figure 2.

In the Wait state, the filter waits for the start signal to go to '1'. After the start signal is high, the filter latches in input data in Get_data state.

The Math state calculates the filter difference equation:

$$Y = b_0 \cdot X + b_1 \cdot X_{k1} + b_2 \cdot X_{k2} + b_3 \cdot X_{k3} + a_1 \cdot Y_{k1} + a_2 \cdot Y_{k2} + a_1 \cdot Y_{k3} \quad (1)$$

The equation is carried out in two steps as shown in Figure 2:

1. Multiply: Each multiply in the filter equation is obtained one at a time by using a single shared multiplier. The data is funneled through the shared multiplier in several clock cycles.

2. Addition: After all the multiplications, the filter goes into addition where it adds the different operands calculated in one step. It uses multiple adders for adding all the operands at the same time.

The output calculated from the filter equation proceeds to the Round/Truncate and Shift state. The output signal is rounded up if it is greater than or equal to 0.5. It is then truncated to get the final output. If the output is less than 0.5, it is truncated to get the final output. The output signal and the input signal both shift down one register for the next iteration:

$$In_{k-1} = In_k$$
$$In_{k-2} = In_{k-1}$$
$$In_{k-3} = In_{k-2}$$

The output signal shifts down in the same manner.

$$Out_{k-1} = Out_k$$
$$Out_{k-2} = Out_{k-1}$$
$$Out_{k-3} = Out_{k-2}$$

If the first two MSBs on the filter output signal go to '01', the error signal goes high and remains high to indicate an overflow in the output. The filter returns to the Wait state and waits for the start signal to go high.

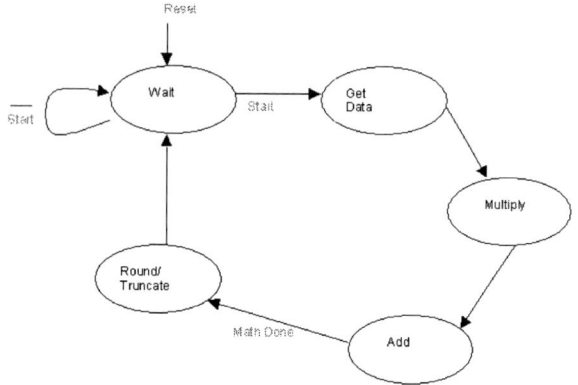

Fig. 2: State machine for Shared Multiplier design

The code uses a total of 18 states to do all the math and truncation. Even though the code uses a large number of flip-flops, this filter is very small in using the space on the FPGA. This method is also time efficient.

III. RESULTS

In simulation the filter matches the expected output exactly. However, we needed to make sure that the filter worked correctly on the FPGA hardware; therefore, we used an analog extension board.

Figure 3 shows the step response of the filter on the FPGA hardware and it is very close to the expected output.

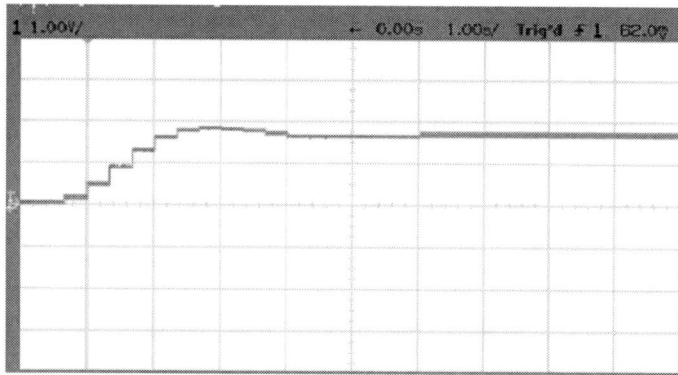

Fig. 3: Actual step response

Figure 4 shows the low pass response for the input signal and demonstrates that it is very close to the expected output.

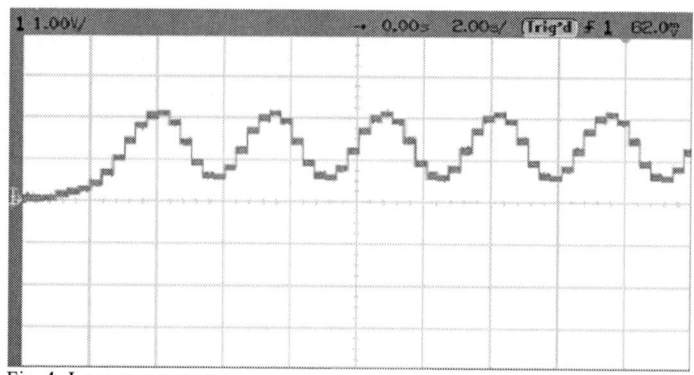

Fig. 4: Low pass response

Figures 3 and 4 show that the filter is working as expected qualitatively on the FPGA hardware. By outputting the digital filter output to a 7-seg display, we verified that the data was also numerically correct.

Table II shows the number of flip-flops, LUTs and the overall space used by the filter.

TABLE II
FLIP FLOP, LUT AND OVERALL SPACE USAGE TABLE FOR THE FILTER DESIGN

	Flip Flop usage	LUT usage	Total space used
Shared Multiplier	5%	8%	11%

The Shared Multiplier design is small in terms of overall space usage (11%) and is also easy to debug and efficient, making it suitable for use in high-temperature FPGA applications.

The design used 3100 gates when the shared multiplier design was implemented onto the high-temperature FPGA for Sandia.

IV. CONCLUSION

The paper outlines the Shared Multiplier design of a space-efficient digital filter for use in High-Temperature FPGA applications.

In terms of overall space usage, the Shared Multiplier design is the very efficient. The Shared Multiplier design is an efficient way of coding the digital filter. The design is space and time efficient.

V. ACKNOWLEDGMENT

The authors acknowledge Sandia National Laboratories for funding this project.

The authors gratefully acknowledge the contributions of Linnea Anderson and Jason Oman from the University of Idaho, and their assistance on the project.

VI. REFERENCES

[1] Sandia National Laboratories, HT83SNL00, www.sandia.gov, 10 February 2006
[2] Digilent, D2FT FPGA www.digilentinc.com, 10 February 2006
[3] Xilinx, Xilinx ISE 7.1i www.xilinx.com, 10 February 2006

40 Gbps SiGe Pattern Generator IC with Variable Clock Skew and Output Levels

Matthew J. Zahller and George S. La Rue, Member IEEE
Washington State University, Pullman, WA

Abstract—**A single-chip 40 Gbps pattern generator design in 0.18 μm SiGe BiCMOS technology is described. An on-chip 128x128 bit RAM with an access time of 3 ns stores the data pattern. A hybrid 128:1 CMOS/ECL multiplexer increases the output data rate from the RAM to 40 Gbps. The output driver is back terminated with 50 ohms and provides programmable levels in the range -2 V to 2 V into a 50 ohm load. The pattern dependent jitter is under 1 ps at all output levels. The clock can be delayed by a programmable number of clock cycles plus a vernier delay of up to 50 ps in 0.2 ps steps. Power dissipation is up to 550 mW depending on the output amplitude and termination voltage.**

Keywords-pattern generator; word generator

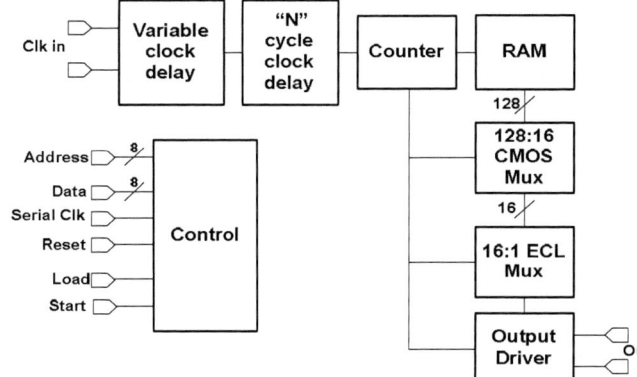

Fig. 1 Block diagram of the pattern generator IC

I. INTRODUCTION

The continued advances in CMOS and bipolar integrated circuit technologies has given rise to higher speed circuits and the need for low-cost high-speed pattern generators to test these circuits. Most high-speed generator ICs only provide pseudo-random bit sequences (PRBS) required for testing communication channels [1-3]. Digital generators with arbitrary patterns are needed to provide stimulus for other circuits, such as multi-GHz digital-to-analog converters. High speed SiGe BiCMOS technology, which combines very high speed heterojunction bipolar transistors (HBTs) and high density CMOS, is a natural choice for integrating pattern memory with high-speed multiplexers and output drivers. InP and GaAs components at 40 Gbps [4] cannot provide one-chip solutions and cost is much higher with these technologies.

The requirements for the high-speed pattern generator are 1) differential outputs with 50 ohm output impedance; 2) programmable output high and low levels; 3) a maximum swing of 1.5 V into 50 ohm loads; 4) programmable delay with resolution of 0.2 ps; 5) a memory depth of at least 64K bits; and 6) pattern dependent jitter less than 1 ps.

II. ARCHITECTURE AND DESIGN

A. Architecture

Fig. 1 shows the block diagram of the pattern generator IC. The pattern data is stored in a 128 x 128 bit SRAM with an access time of 3 ns. The 128-bit wide output words are first multiplexed by sixteen 8:1 CMOS multiplexers to 2.5 Gbps and then by a 16:1 HBT ECL multiplexer to 40 Gbps. The final multiplexer uses the clock to directly select the data thus,

reducing the maximum clock rate to 20 GHz. This puts a requirement on the duty cycle to be 50% however. The high rate data is then fed into the output driver which contains a variable level shift circuit, three DACs to control the variable output high and low levels, and an output differential pair.

The variable delay is achieved by delaying the clock in two separate delay circuits. The vernier delay provides a programmable delay of up to 50 ps in 0.2 ps steps. The second delay circuit delays the clock by a programmable number of up to 255 clock cycles. The counter that addresses the RAM and multiplexers uses a combination of SiGe bipolar and CMOS logic. The high speed clock is first divided using bipolar flip-flops and then converted to CMOS levels when speeds are low enough. In order to provide lower data rates without duplicating data in the memory, the CMOS clock that drives the SRAM column counter can be divided by 1, 2, 4, or 8.

The SRAM is organized as 128 columns of 128-bit words. Since the output word width of the SRAM is large, the 8-bit counter that addresses the columns is only required to operate at 300 MHz. The counter can perform looping to effectively increase the pattern length.

B. Design

Fig. 2 shows a block diagram of the output driver. The output driver consists of an initial level shift circuit that controls the output amplitude of the data driving the final output differential pair. The variable level shift circuit is utilized to change the amplitude of the incoming signal provided to the output differential pair in order to reduce the output jitter of the data. Four 8-bit DACs control the output driver. One DAC controls the level shift differential pair bias, another controls the output data high level and a third controls

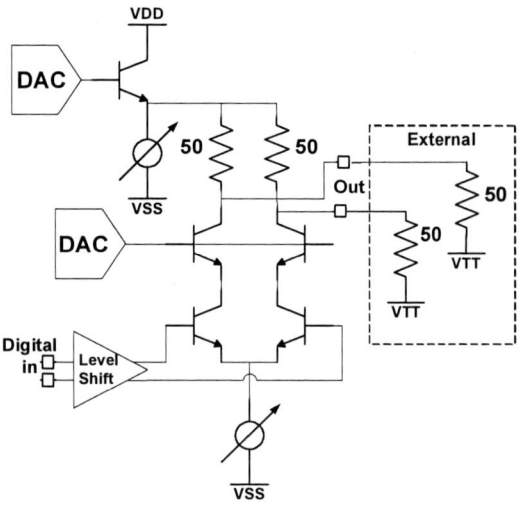

Fig. 2. Block diagram of the output driver

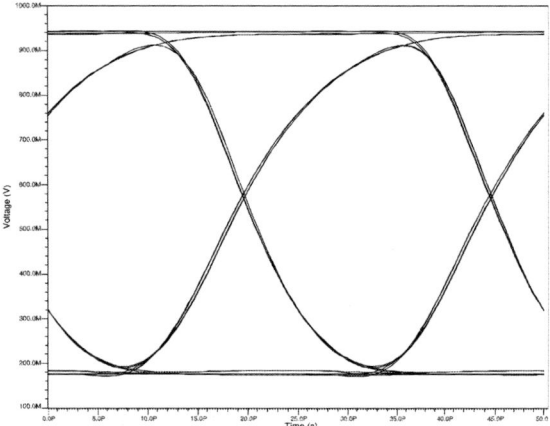

Fig. 3. Simulated eye diagram of the output at 40 Gbps and an output amplitude of 0.75V.

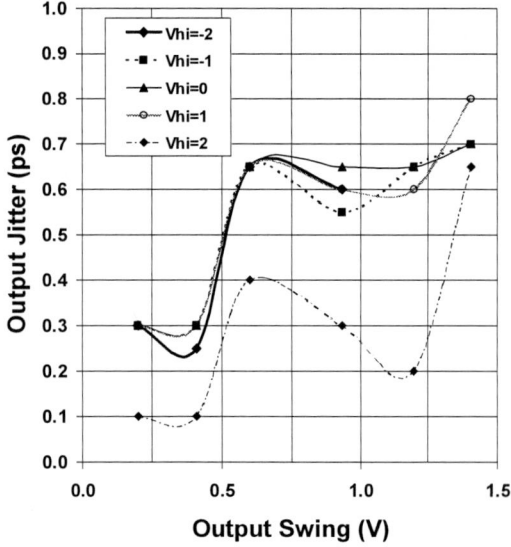

Fig. 4. Simulated data dependent output jitter as a function of output amplitude at different high output levels.

the current to the output differential pair to control the output amplitude. Depending on what the high level is and what termination voltage is used, a fourth DAC controls a current sink to adjust the high output level. A coarse DAC is used to control the voltage of the cascode HBTs in the output stage to remain below the breakdown voltage. Fig. 3 shows a simulated eye diagram of the output at 40 Gbps with an amplitude of 0.75 V. The pattern dependent jitter is less than 1 ps. Fig. 4 plots the output jitter as a function of output amplitude for different output high levels. The worst case jitter is 0.8 ps. The total output jitter will depend on the jitter of the input clock and any jitter added by the vernier delay and clock distribution circuits.

The programmable vernier delay of zero to 50 ps was realized using two zero to 8 ps delay circuits as well as multiple inverter chains and multiplexers. The 8 ps variable delay circuit uses current steering with three ECL inverters [2]. The configuration of the inverters allow for a signal path consisting of one to two inverters. By varying the current between the two paths the delay can be varied from 0 to 8 ps.

The second delay circuit utilizes an 8-bit ECL counter that is preset to the cycle delay value before starting. The counter begins to count down after the start is enabled and when it reaches zero a flip-flop is set to allow the main clock signal to pass through to the counter circuit. This circuit delays the output data by any number of cycles up to 255 and reduces the maximum delay required for the vernier.

III. CONCLUSION

A 40 Gbps pattern generator design is described in 0.18 µm SiGe BiCMOS technology. An on-chip 64 K-bit RAM with an access time of 3 ns stores the data pattern. A hybrid 128:1 CMOS/ECL multiplexer increases the output data rate to 40 Gbps. The output driver is back terminated with 50 ohms and provides programmable levels in the range -2 V to 2 V into a 50 ohm load. Simulations show that the pattern dependent jitter is under 1 ps at a wide range of output levels. The clock can be delayed by a programmable number of clock cycles plus a vernier delay of up to 50 ps in 0.2 ps steps. Power dissipation is up to 550 mW depending on the output amplitude and termination voltage.

REFERENCES

[1] D. Kucharski and K. Kornegay, "A 40Gb/s 2.5V 2^7-1 PRBS generator in SiGe using a low-voltage logic family," *IEEE Int. Solid-State Circuits Conf. Dig. Tech. Papers*, pp. 340-341, 602, 2005.

[2] T. Dickson, et al., "A 72 Gb/s 2^{31}-1 PRBS Generator in SiGe BiCMOS technology," ," *IEEE Int. Solid-State Circuits Conf. Dig. Tech. Papers*, pp. 342-343, 602, 2005.

[3] M.G. Chen and J.K. Notthoff, "A 3.3-V 21-Gb/s PRBS generator in AlGaAs/GaAs HBT technology," *IEEE J. Solid-State Circuits*, vol. 35, pp. 1266-1270, Sept. 2000.

[4] D. Streit, et al., "InP and GaAs components for 40 Gbps applications," *GaAs IC Symposium*, pp. 247-250, 2001.

High Speed Digital Input Buffer Circuits

Krishna Duvvada, Vishal Saxena and R. Jacob Baker
ECE Dept., Boise State University, krishnaduvvada@boisestate.edu

Abstract—**This paper illustrates design, fabrication and testing of novel differential high-speed digital input buffers. The delay of the proposed input buffers are nearly independent of power supply voltage and input signal amplitudes. The pulse shape of the output signal is highly symmetric which mitigates skew related errors.**

Keywords- CMOS, Differential Amplifier, Digital design, High-frequency Input Buffer.

I.   INTRODUCTION

Input buffer circuits are present at a chip's input and convert input signals with imperfections such as slow rise and fall times into clean, full logic level digital signals for use inside the chip. If the buffer doesn't slice the data at the correct time instants, timing errors can occur. If the input signal is sliced too high or too low, the output signal's width is incorrect. In the high speed systems this reduces the timing budget in the systems and can result in errors [1]. This paper presents design, fabrication and test results of novel, differential high-speed input buffers which mitigate these problems.

II.   DIFFERENTIAL INPUT BUFFER CIRCUITS

In order to precisely 'slice' the input data, the data is transmitted differentially as an input and its complement. A differential amplifier input buffer amplifies the different between the two inputs. The buffer topologies used in this design employ self biased differential amplifiers as no external reference is used to set the bias current in the diff-amp [1],[2]. Fig.1 shows an NMOS version of input buffer. When the input falls below V_{THN}, then the circuit will not work very quickly as the NMOS are moved into subthreshold region. This will result in an increase in the delay [1].

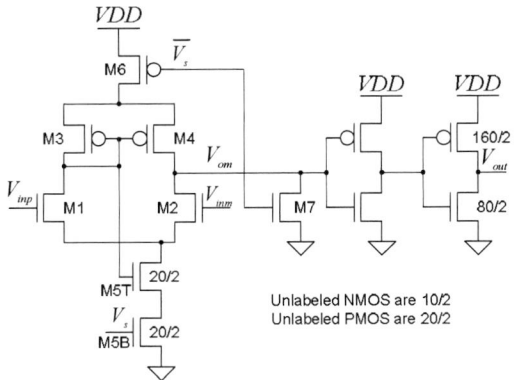

Figure 1.   NMOS input buffer

Ideally, the delay of the buffer should be independent of power supply voltage, temperature, input signal amplitudes or pulse shape. In order to obtain better performance for lower input level signals, a PMOS version of input buffer (fig.2) can be used. However this scheme leads to the appearance of a large offset. To avoid the offset, the NMOS buffer can be used in parallel with a PMOS buffer as shown in fig.3, to form an input buffer that operates well with input signals approaching ground or *VDD*. The topology with the buffers in parallel provides a robust input buffer that works for a wide range of input voltages [1].

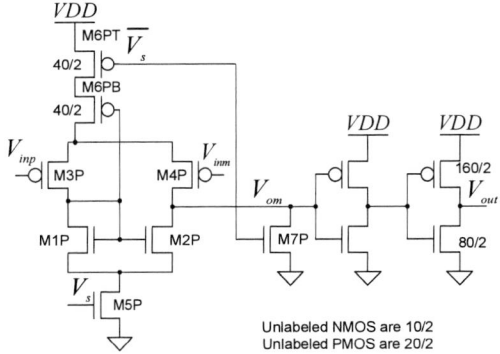

Figure 2.   PMOS input buffer

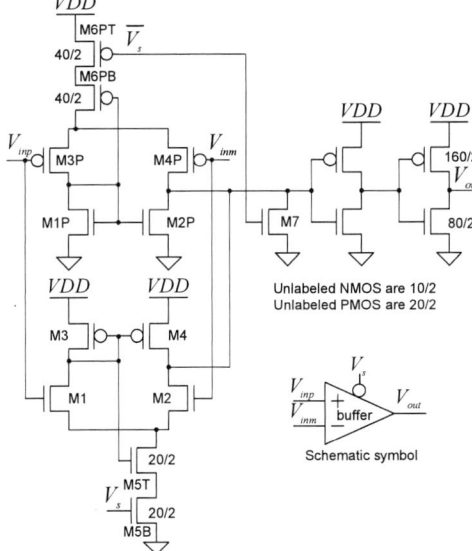

Figure 3.   Rail-to-rail input buffer combining NMOS and PMOS buffers

The input buffers are designed with 'enable' logic using the NMOS and PMOS switches, like M5B and M6 in fig.1, controlled by signals V_s and $\overline{V_s}$. When the enable signals are off, the output of the diff-amp will be in high impedance state. This can cause a large amount of current to flow in the inverters and damage the chip. To avoid this scenario, a switch M7 is connected to the output of the diff-amp which clips the node to ground when the buffer is disabled.

The buffers are fabricated in AMI's CN5 (0.5μm) process with a *VDD* of 5V as shown in fig. 4. The buffers are designed to drive a load of 30pF which accounts for the load offered by the probe setup. A large inverter (160/80) is used to drive the large load with optimal delay.

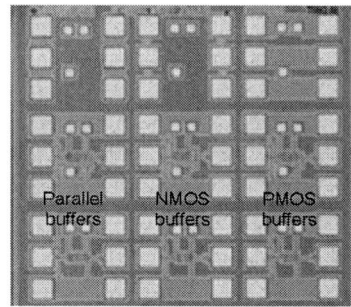

Figure 4. Micrograph showing the input buffer circuits fabricated on a chip.

III. TEST RESULTS

The input buffers were tested using the setup shown in fig.5. The common mode reference voltage V_{ref} was nominally kept at 2.5V and the input differential signal v_p (clock pulses with 1MHz frequency) was applied at the positive terminal of the buffer. The net delay was measured as the sum of charging and discharging delays (i.e. $t_{pLH} + t_{pHL}$). Figs.6-9 show the net delay offered by the input buffers for varying supply voltage, common mode reference voltage, input signal swing and temperature respectively.

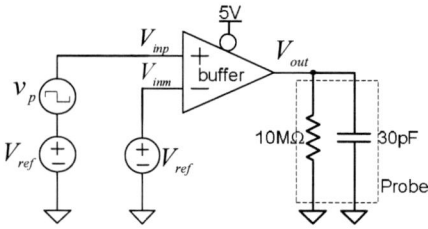

Figure 5. Test setup for the input buffers

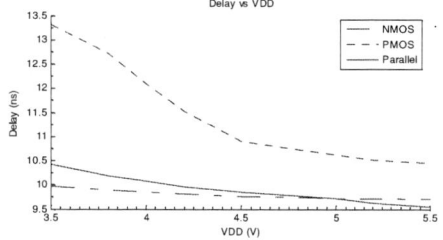

Figure 6. Delay vs supply voltage (*VDD*) plot for the input buffers.

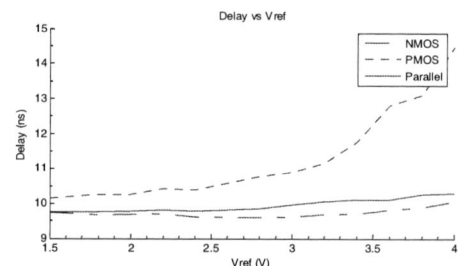

Figure 7. Delay vs common mode reference (*Vref*) plot for the input buffers.

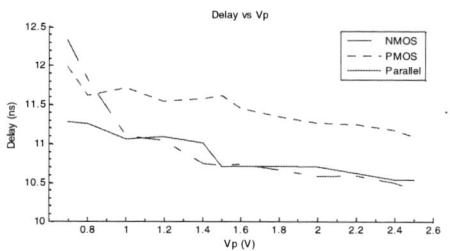

Figure 8. Delay vs input signal swing (*Vp*) plot for the input buffers.

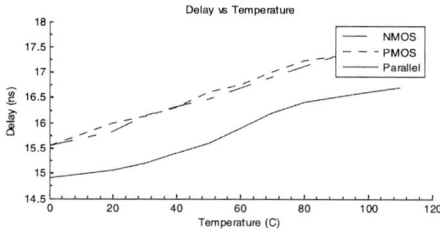

Figure 9. Simulated delay vs temperature plot for the input buffers.

The test results demonstrate that the designed input buffers operate well for high-speed input signals. The delay is virtually independent of power supply voltage, input common mode reference and voltage swing. The output pulse is highly symmetric and skew is absent. The parallel buffer topology provides the optimal performance for wide range of input voltages.

IV. CONCLUSION

A set of differential high-speed input buffers have been designed, fabricated and tested. The designed input buffers provide high frequency operation with lower delays. The delays are nearly independent of variations in power supply, input signal common mode and differential voltages, and temperature.

REFERENCES

[1] R. J. Baker, *CMOS: Circuit Design, Layout and Simulation, 2nd ed.* Boise, ID: Wiley-IEEE, 2005, pp. 531-538

[2] M. Bazes, "Two Novel Fully Complementary Self-Biased CMOS Differential Amplifiers," *IEEE Journal of Solid State Circuits*, vol. 26, no. 2, Feb. 1991.

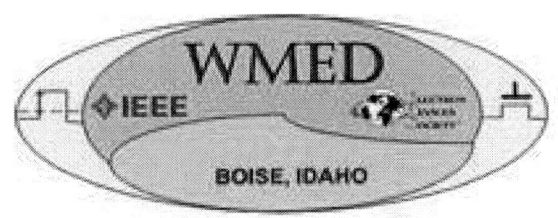

Session

MEMS, Devices, and Sensors

Polymer based thin film coils as a power module of wireless neural interfaces

S. Kim[1], K. Buschick[2], K. Zoschke[2], M. Klein[2], M. Toepper[2], D. Black[1], R. Harrison[1], P. Tathireddy[1], F. Solzbacher[1]

[1]Dept of Electrical and Computer Engineering, University of Utah, Salt Lake City, UT, USA
[2]Fraunhofer IZM / Technical University of Berlin, Berlin, Germany

Abstract – **For the conventional Utah Electrode Array (UEA) to be able to function without transcutaneous wire connections, a kind of power source is needed in an integrated form with the UEA. To develop such wireless neural interfaces, inductive coupling between two coils was used to deliver power to the integrated electronics. The power receiver coil was microfabricated as a polymer based component, and its electrical characteristics and performance in power transmission were investigated in dry condition.**

I. INTRODUCTION

Recently, efforts have been devoted to develop a fully integrated, wireless neural recording device based on the conventional Utah Electrode Array. This will free the patient from the risk of infection associated with a wired connection and allow distribution of a network of interface nodes through the central and peripheral nervous system. To this end, fully integrated neural interfaces need to have a wireless power source and the capability to wirelessly transmit data to/from extracorporeal devices. Inductive coupling between two coils can be a solution to provide power and data to the implanted electronics. In this paper, coils serving as a power module to supply the integrated neural interface were fabricated, characterized, and its power transfer performance was tested in laboratory condition.

II. FABRICATION OF POWER COILS

Taking into account simulation results [1], technological considerations, the UEA re-routing strategy, and the device assembly process, several optimized coil designs were manufactured based on polyimide. Single and double layer coils were manufactured with line width and spacing of 15-20 μm and a thickness of larger than 10 μm of the Au layer.

PI/Au coils were fabricated on a 4" Si wafer. The process is based on a release layer which is deposited directly on the monitor wafers. This layer based on non-filled thermoplastic polymer is necessary for the final separation of the thin film PI/metal stack. Each wafer carries 100 coils with different designs. The process for the double layer coils consists of four plating steps, starting with the polymer layer and ending with the formation of Ni/Au interconnection pads. The plating process is done using the combination of sputtering TiW/Au and deposition of photoresist (AZ 9000 from AZ Electronic Materials). The PI is a polyamic acid type from Fuji-Film. The left of Figure 1 shows an optical microscope image of an electroplated Au coil. Six different geometries of Au coils were electroplated on PI substrates. The right of Figure 1 shows the cross section of a 15 μm/15 μm width/space double-layered PI coil as an example of the fabricated coils.

With this width and spacing, 60 turns could be realized on each layer.

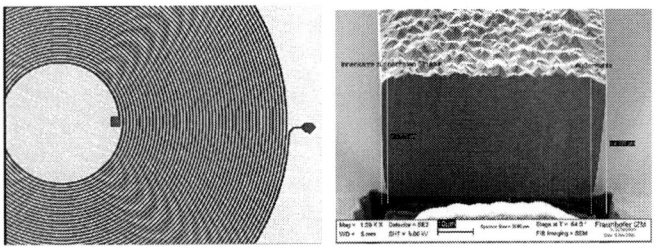

Figure 1: Optical microscope image of an electroplated Au-coil with 15 μm width and 15 μm space on polymer layer (left), FIB cut of Au coil (right).

To avoid undesirable high inter-winding capacitance that may exist between coil layers when using double layer coils, stackable coils were proposed. These can be switched in parallel or series in order to tune the coil parameters and therefore the resonant frequency of the circuit. The coil stack for inductive power coupling consists of two polyimide based electroplated Au coils with 60 turns each. The coils are glued on a low-temperature-co-fired-ceramic (LTCC) ferrite platelet to increase the Q factor. An electrical interconnect layout was designed to connect the PI based Au coil through vertical spacers to the integrated electronics and other surface mount device (SMD) components. The schematic in Figure 2 (left) depicts the routing layout and dimensions of the package. To compensate for varying parasitic capacitances and voltage gain, a jumper can be used to operate the coils in three ways: 1) a double layer coil, 2) a single layer coil, and 3) two stacked coils (a double and a single or two single coils). A mounted coil on a test module is shown in Figure 2 (right).

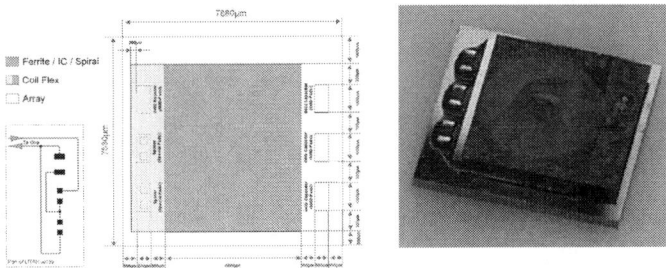

Figure 2: Schematic of the dimensions of the package; the two spacers and the jumper on the left side can be used for single or double coil assembly, the circuit plan is shown on the bottom left side (left), light microscopy image of a fully assembled test module with mounted coil/ferrite (right).

III. CHARACTERIZATION AND TEST OF COILS

All six geometries of the polyimide based coils were tested with and without the LTCC ferrite platelet backing. The coils were mounted and wire bonded on standard DIP40 ceramic packages for convenience in testing. The coils were mounted outside the cavity of the packages to avoid eddy currents associated with the conductive substrate in the cavity.

Inductance and resistance measurements were taken using a Stanford Research Systems SR715 LCR meter at 10 kHz. The results are shown in Table I. Because inductance and resistance are somewhat frequency dependant, the Q factor was measured empirically at the resonant frequency of 2.64 MHz. The method for this measurement is described in more detail below. Coil type A in Table I has 51 windings with a width of 20 μm and space between windings of 15 μm. Type B also has 51 turns, but with width/spacing of 15 μm/20 μm, and type C has 60 turns with width/spacing of 15 μm/15 μm. All three types have a diameter of 5 mm. Single and double layer versions of these make up the six geometries investigated. Each coil type was tested with and without ferrite backing, resulting in twelve total coil configurations.

For the same number of turns, coils of type B have more series resistance than those of type A. This is due to the smaller cross sectional area of the windings. Type C coils have more turns and thus have greater inductance and greater resistance. Double layer coils had approximately four times the inductance of similar single layer coils due to having double the number of turns. The addition of the ferrite backing resulted in an average increase in inductance of 64 % without significantly affecting resistance. This led to an average increase in the Q factor of 82 %.

The Q factor was measured at resonance using the relation $Q = f_0/BW$, where f_0 is the resonant frequency of 2.64 MHz and BW is the -3 dB bandwidth of the coil. This was accomplished by first selecting appropriate capacitors for each coil to resonate at f_0. These capacitors were chosen empirically to avoid discrepancies due to parasitic capacitances in the test setup. A frequency sweep was then performed to determine the -3 dB point on either side of the peaking. BW is the difference between the two -3 dB frequencies. The frequency sweep was performed using an Agilent 33120A function generator with 10 V_{pp} amplitude. This supply has a fixed output source resistance of 50 Ω, which is comparable to the series resistance of the coils. Therefore, a 1 MΩ resistor was inserted in series with the coil to minimize the potential distortion of data due to frequency-dependant changes in coil inductance and resistance.

Next, each coil was tested using resonant capacitors to supply power to the integrated electronics (currently packaged separately) [2]. The combined load of these electronics was about 3 mA. It was found that higher inductance and higher Q coils were most effective for powering the electronics. Approximately 4.5 V is required at the receive end to provide suitable headroom for a rectified, regulated 3.3 VDC power supply. All three double-layer, ferrite-backed coils were able to supply sufficient voltage to power the electronics. The type A coil was the most efficient, able to provide the necessary 4.5 V at a distance of up to 12 mm.

TABLE I.   ELECTRICAL MEASUREMENTS

Coil Configuration			Measurements		
			L (μH)	R (Ω)	Q @ 2.64 MHz
Single Layer	No Ferrite	A[a]	7.29	33.5	2.49
		B	7.41	40.5	2.18
		C	10.3	50.5	2.78
	With Ferrite	A	11.3	36.7	4.98
		B	12.1	40.0	4.80
		C	16.9	50.5	5.08
Double Layer	No Ferrite	A	29.2	74.8	5.62
		B	29.1	88.6	4.98
		C	40.2	111.3	5.18
	With Ferrite	A	46.9	74.5	9.10
		B	49.8	88.5	8.25
		C	68.5	113.2	8.52

a. Coil Type A has 51 turns, winding width of 20 μm, and space between windings of 15 μm. Type B has 51 turns, width/spacing of 15 μm/20 μm; type C has 60 turns with 15 μm/15 μm.

IV. CONCLUSIONS

In this work, power coils to supply power to the integrated wireless neural interface were fabricated and characterized, and its power transmission performance was fully tested in laboratory condition. The fabricated Au thin film coils based on polyimide can be switched between single layer, double layer, or two stacked coils with the help of a properly designed packaging layout. The most efficient coil type, a double layer coil with 51 turns, width/spacing of 20 μm/15 μm, and a ferrite plate underneath showed an inductance of 46.9 μH and Q factor of 9.10 at 2.64 MHz, resulting in sufficient power supply to the electronics, e.g. greater than 4.5 V over a distance of up to 12 mm.

ACKNOWLEDGMENTS

This work was supported by NIH contract No. HHSN265200423621C.

REFERENCES

1. Rieth, L. et al., "Switchable LTCC/Polyimide Based Thin Film Coils," presented in *2005 NIH/NINDS Neural Interfaces Workshop*, Bethesda, MD, Sep 2005.
2. Harrison, R. et al., "A low-power integrated circuit for a wireless 100-electrode neural recording system," In: *IEEE Intl. Solid-State Circuits Conf. (ISSCC 2006) Digest of Technical Papers*, pp. 554-555, San Francisco, CA, 2006.

Design and Fabrication of a MEMS Capacitive Chemical Sensor System

Vishal Saxena, Todd J. Plum, Jeff R. Jessing and R. Jacob Baker

ECE Dept., Boise State University, vishalsaxena@ieee.org

Abstract— **This paper describes the development of a MEMS sensor system to detect volatile compounds. The sensor consists of a MEMS capacitive sensor element monolithically integrated with a sensing circuit. The sensor element is a parallel plate capacitor using a chemically sensitive polymer as the dielectric. In presence of the target analyte, the polymer swells and changes the capacitance of the sensor element. This change in capacitance is sensed and converted to a digital bit stream by a delta-sigma sensing circuit. This paper provides an overview of the design of the sensor element, the sensing circuit and the process integration for their fabrication on a single die.**

Keywords-MEMS Chemical Sensors, Chemicapacitive Sensors, Delta-sigma Sensing, Fabrication, Process Integration.

I. INTRODUCTION

Chemical sensors are being widely deployed for environmental monitoring, industrial hazard detection and sensing of chemical warfare agents. Micro-electro-mechanical systems (MEMS) technology has enabled design of miniaturized and fully integrated chemical sensors. MEMS chemical sensors detect presence of chemical analytes and produce electrical responses proportional to the concentration of the target analyte. These electrical responses are detected and amplified using a sensing circuit which is typically fabricated using CMOS (Complementary Metal Oxide Semiconductor) process. Monolithic integration of MEMS sensors and interface circuits results in a low power, autonomous, miniaturized, and reliable sensor system-on-a-chip (SoC) [1]. This sensor SoC can be easily extended to include an RF subsystem to construct self-sustained wireless sensor networks [2].

This paper presents the development of a MEMS capacitive chemical sensor system. The following sections describe the sensor system architecture, design of the MEMS sensor element and sensing circuit, and their fabrication.

II. CAPACITIVE MEMS SENSOR SYSTEM

The capacitive MEMS sensor system (fig.1) detects volatile analytes by producing a change in capacitance, proportional to the target analyte concentration.

The sensing circuit converts the change in capacitance to a voltage signal. The sensed voltage is converted to a digital signal, which is output to an external logic block for data processing. The MEMS sensor element and the sensing circuit are both to be integrated on the same die.

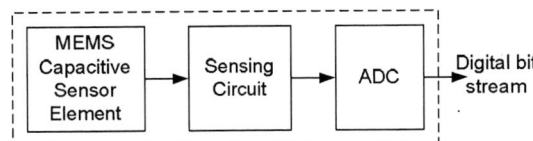

Figure 1. Block diagram of the chemical sensor system.

III. MEMS SENSOR ELEMENT

The sensor element is designed as a parallel-plate capacitor composed of overlapping metal layers with a chemically sensitive polymer used as the dielectric (fig.2). Neglecting the fringe capacitance, the sensor capacitance is given as $C_s = \varepsilon_{poly} A/t$, where ε_{poly} is the net dielectric permittivity of the polymer, 't' is the mean polymer thickness and 'A' is the metal plates overlap area.

When the sensor is exposed to the target analyte, absorption from the gaseous phase into the bulk of polymer takes place. It results in swelling in the polymer (i.e. t increases) and an increase in dielectric permittivity (ε_{poly}). These changes in sensor property change the sensor capacitance as $\delta C_s / C_s = \delta\varepsilon/\varepsilon - \delta t/t$ [1].

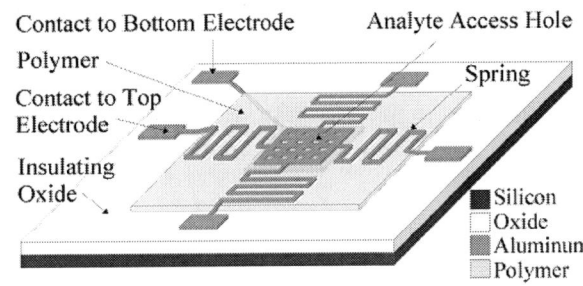

Figure 2. A MEMS capacitive sensor element [3].

The effects of polymer swelling and change in permittivity, on the change in the sensor capacitance, counter each other and may mutually cancel. To avoid this situation the polymer is carefully chosen such that the polymer swelling is maximized for the target analyte. The permittivity of the selected polymer should be as small as possible for maximum sensitivity. The capacitive sensor designed in this work relies upon the polymer swelling as the dominant effect. The polymer which is being experimented in the first sensor prototype is poly-ethylene vinyl-acetate (PEVA) which swells on exposure to benzene [3].

Support of this project by EPA Contract No. X-97031101-0 is gratefully acknowledged.

IV. Sensor Interface Circuit

For circuit design, the capacitive MEMS sensor element is modeled as a variable capacitor. A MEMS capacitive sensor design is reported in [4], which utilizes the change in permittivity of the polymer dielectric upon exposure, to sense the target analyte. The base capacitance of this sensor is around 10pF and the incremental change in capacitance is 4aF/ppm. In order to sense such small variations, a high resolution and noise tolerant sensing circuit is required.

A delta-sigma modulation based sensing circuit topology is ideal for this application as it provides high-resolution, noise tolerance, high linearity and compact circuit integration. The advantage of employing delta-sigma sensing is that the analog to digital converter (ADC) is inherently realized in the topology [4].

Fig.3 illustrates a simple first order delta-sigma (over-sampling) sensing topology which uses feedback to balance the charge flows from the sensor capacitor C_s and reference capacitor C_{ref}, to capacitor C_f. The circuit uses non-overlapping clocks φ_1 and φ_2. The reference capacitor is identical to the sensor capacitor, except that is hermetically sealed to avoid exposure to the analyte. The reference capacitor is used for differential sensing which eliminates the effects of aging i.e. change in sensor properties with time and usage [5], [6].

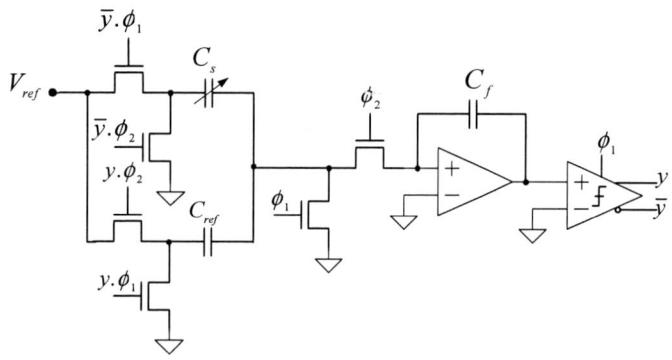

Figure 3.   A first order delta-sigma sensing circuit.

The estimate of normalized change in capacitance is given as $(C_s - C_{ref})/(C_s + C_{ref}) = 1 - 2\hat{y}$, where $\hat{y}$ is the average of the output binary data stream, $y[n]$ [6]. At first, the sensor circuit is designed using only n-type MOSFETs (NMOS) so that it can be fabricated in-house using available NMOS process [7] and easily integrated with the capacitive MEMS sensor.

V. Fabrication Process Integration

The first sensor prototype is fabricated in the back end of the line (BEOL) of a simple metal-gate NMOS fabrication process, developed at BSU cleanroom [7]. To start with the process, a 5000Å thick sacrificial oxide is grown on a (100) p-type wafer. This oxide is patterned with 'active' mask to create drain source regions. The drain source regions are doped with phosphorus using spin-on dopant film, followed by a drive-in step. The sacrificial oxide is stripped and 5000Å thick oxide is

grown. This oxide is patterned with 'gate' mask to create gate region windows which are aligned with the source/drains. Then a high quality 200Å gate oxide is grown using dry oxidation.

The wafer is now patterned using the 'contact' mask to etch contact holes into the field oxide. Then, a 2500Å aluminum film is sputtered to fill the contact holes and deposit metal1 layer. This metal layer is pattern with 'metal1' mask to create bottom electrode of the sensor and wires for NMOS circuits. After this, an oxide layer is blanket deposited which is patterned with 'via1' mask to open via holes and bottom plate of the sensor element. Next, the chemically sensitive polymer is applied using a spin coater to yield a uniform, 1μm thick film. A 'titanium etch' mask is sputtered and patterned on the polymer. The polymer is then etched in a barrel asher using O_2 and CF_4 as reagent gases. Finally a 2500Å aluminum film is sputtered with 'metal2' mask to create the top electrode of the sensor.

VI. Progress and Future Work

MEMS Sensor elements are currently being fabricated. A simple NMOS process for fabrication of sensing circuits has been developed. Presently delta-sigma sensing topologies are being designed, which will be fabricated and tested. Finally the MEMS sensor element will be integrated with the sensing circuit on a single die and tested.

VII. Conclusion

A MEMS capacitive sensor system for sensing volatile chemicals has been proposed. The design and fabrication method for the sensor element and a simple sensing circuit is presented. The MEMS sensor elements are being fabricated. A simple in-house NMOS process is developed for fabrication of sensor interface circuits. After isolated testing of the sensor element and the sensing circuit, both will be monolithically integrated on a single die.

References

[1] A. Hierlmann, *Integrated Chemical Microsensor Systems in CMOS Technology*, Zurich, Switzerland: Springer-Verlag, 2005.

[2] A. Arora et al, "A Wireless Sensor Network for for Target Detection, Classification, and Tracking," Computer Networks Journal, Oct. 2004.

[3] T. J. Plum, V. Saxena and J.R. Jessing, "Design of a MEMS Capacitive Chemical Sensor Based on Polymer Swelling," *submitted to IEEE WMED 2006.*

[4] C. Hagleitner, A. Heirlemann, and H. Baltes, "Single-Chip CMOS Capacitive Gas Sensor for Detection of Volatile Organic Compounds," *IEEE Sensors Conference 2002*, vol. 2, pp. 1428-1431.

[5] R. J. Baker, *CMOS: Circuit Design, Layout and Simulation, 2nd ed.* Boise, ID: Wiley-IEEE, 2005, chap. 25.

[6] B. Wang, T. Kajita, T. Sun and G. Temes, "High-Accuracy Circuits for On-Chip Capacitive Ratio Testing and Sensor Readout," *IEEE Tran. On Instr. And Meas.*, vol. 47, No. 1, Feb. 1998.

[7] NMOS Fabrication Process, Idaho Microfabrication Laboratory, Boise State University, [Online]. Available:

http://coen.boisestate.edu/imfl/Processes/NMOS_Process.doc

A High Sensitive Piezoresistive Sensor for Stress Measurements in Packaged Semiconductor Die

Ahsan Mian, Jeffrey C. Suhling, and Richard C. Jaeger

Abstract— **The authors have developed new two-dimensional piezoresistive stress sensors that replace conventional serpentine resistor rosettes. These sensors are named van der Pauw (VDP) sensors as they are based upon four-terminal van der Pauw type resistance measurements. The resistance of such a sensor is size independent, and hence can be made as small as lithographically possible to capture stresses in critical areas on the surface of a packaged semiconductor die. It was predicted theoretically that the VDP sensor should exhibit a greater than three times improvement in sensitivity relative to resistor sensor rosettes. Then the response of actual VDP structures fabricated on (111) silicon surface was characterized under uniaxial load using four-point-bending tests. These experimental results confirm that the VDP stress sensitivities are more than three times higher than those of their corresponding resistor sensor counterparts.**

I. INTRODUCTION

For many years, resistive stress sensors have been used successfully to measure die stresses in a wide range of packaging applications [1-2]. However, resistor sensors possess several drawbacks. Diffused and implanted resistors have high temperature sensitivity relative to the stress response, and great care must be exercised to achieve accurate measurement of stress. Resistors are often designed with relatively large meandering patterns to increase the total resistance, but they then suffer from transverse sensitivity which is difficult to estimate due to the lateral diffusion that occurs during the fabrication process. Transverse sensitivity can be minimized by interconnecting resistor legs with metal links, but these require additional contacts that further increase the resistor size.

In a paper published recently [3], the authors have fully developed the theory for a new piezoresistive stress sensor termed the van der Pauw (VDP) sensor that is designed to replace conventional serpentine resistor sensors. The sensor utilizes the technique of measuring "resistances" based on the theoretical developments of van der Pauw. When used as stress sensors, VDP devices have the potential to reduce some of the error sources in resistor-based sensors. As will be demonstrated, they offer a greater than thee times increase in sensitivity compared to that of resistors, and therefore they also exhibit a corresponding reduction in

A. Mian is with the Dept. of Mechanical and Industrial Eng., Montana State University, Bozeman, Montana 59717 (amian@me.montana.edu)

J. C. Suhling is with the Dept. of Mechanical Eng., Auburn University, Auburn, Alabama 36849

R. C. Jaeger is with the Dept. of Electrical and Computer Eng., Auburn University, Auburn, Alabama 36849

sensitivity to thermal errors. The classical VDP structure itself requires only one square of material plus room for four contacts. Thus, these sensors have the potential to be made small enough to capture localized stress variations without any loss of sensitivity. The VDP characteristics are size independent and do not have the transverse sensitivity issue. In contrast, it is the distributed nature of the VDP structure that leads to the enhanced sensitivity.

In this work, the sensitivity of the VDP sensor is first predicted theoretically. Then, VDP and resistor sensors fabricated on (111) silicon surfaces are characterized using uniaxial stress, and the predicted enhancement of the sensitivity of the sensors is experimentally demonstrated.

II. VDP SENSOR PIEZORESISTIVE THEORY

The van der Pauw sensor has four electrical contacts A, B, C, and D located at the four corners of the sample as shown in Fig. 1. The "resistance" $R_{AB,CD}$ of the sample is the potential difference $V_D - V_C$ between contacts D and C per unit current through contacts A and B. A similar resistance $R_{BC,DA}$ can be defined in an analogous manner. The orientation ϕ of the sensor indicates the angle between the x_1'- axis and the line connecting points A and B. Also, a simplified notation is introduced for the resistance of the oriented VDP sensor:

$$R_\phi = R_{AB,CD} = (V_D - V_C)/I_{AB} \qquad (1)$$

Note that a given VDP structure as shown in Fig. 1 can be used to measure both R_ϕ and $R_{\phi+90}$.

A local coordinate system has also been considered denoted by the $x_1'' - x_2''$ axes that are directed along the edges of the VDP sensor. As developed in [3], the sensor resistance change equations for such sensors fabricated on (111) silicon surfaces are given as

$$\frac{\Delta R_0}{R_0} - \frac{\Delta R_{90}}{R_{90}} = 3.16 \begin{bmatrix} (B_1 - B_2)\sigma_{11}' - (B_1 - B_1)\sigma_{22}' \\ + 4\sqrt{2}(B_2^n - B_3^n)\sigma_{23}' \end{bmatrix} \qquad (2)$$

$$\frac{\Delta R_{45}}{R_{45}} - \frac{\Delta R_{-45}}{R_{-45}} = 3.16\left[4\sqrt{2}(B_1 - B_2)\sigma_{13}' + 2(B_1 - B_2)\sigma_{12}'\right] \qquad (3)$$

The normalized resistance change is defined by

$$\Delta R_\phi / R_\phi = (R_\phi^\sigma - R_\phi^0)/R_\phi^0 \qquad (4)$$

where R_ϕ^σ is the resistance measured for the stressed VDP, and R_ϕ^0 is the reference resistance measured for the "unstressed" sensor.

In order to verify the theory, we focus on the uniaxial stress case where $\sigma_{11}' = \sigma$ is applied in x_1' direction, and all other stress components are assumed to be zero. For this case, the difference between the normalized resistance changes for a $\phi = 0^{\circ}$ VDP sensor can be obtained from Eqs. (2) and (3) as

$$\frac{\Delta R_0}{R_0} - \frac{\Delta R_{90}}{R_{90}} = 3.16(B_1 - B_2)\sigma \qquad (5)$$

whereas the difference between the normalized resistance changes for a $0/90^{\circ}$ pair of resistors calculated in [2] is

$$\frac{\Delta R_0}{R_0} - \frac{\Delta R_{90}}{R_{90}} = (B_1 - B_2)\sigma \qquad (6)$$

The responses in Eq. (5) is 3.16 times larger than those produced by the corresponding resistor rosette fabricated on the same wafer with the same doping level and result from the distributed nature of the two-dimensional VDP sensor. In the next sections, this enhanced sensitivity is verified experimentally.

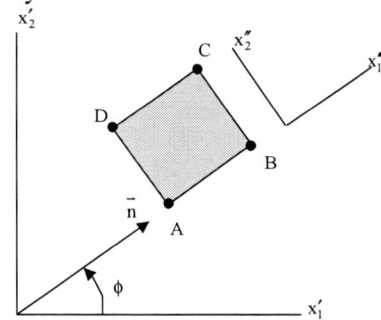

Figure 1 - Oriented Rectangular VDP Sensor

III. EXPERIMENTAL MEASUREMENTS

For experimental verification, VDP test structures were fabricated on (111) silicon surfaces using standard semiconductor processing at Auburn University. As an example, Fig. 2 shows the layout of the BMW-2 resistor test chip, which has been fabricated using (111) silicon wafer. In this die, a separate VDP test cell was included which contains two p-type and two n-type VDP sensors. For each type of doping, one VDP sensor is oriented at $\phi = 0^{\circ}$ (allowing measurement of R_0 and R_{90}), and the other is oriented at $\phi = 45^{\circ}$ (allowing measurement of R_{45} and R_{-45}). In the present analysis, only results for the $\phi = 0^{\circ}$ sensors are discussed, since the 45° sensors are designed to measure shear stress which is not easily applied in a controlled manner. The test chip also contains the resistor rosette test site indicated in Fig. 2 as well as process test sites and a dozen measurement rosettes.

The variation of the resistances of the VDP structures with applied uniaxial stress has been measured using four-point bending tests. The silicon wafers containing the test chips were cut into rectangular strips. Each strip contains a series of chips, and was loaded in a four-point bending fixture to apply uniaxial stress. Resistances R_0 and R_{90} were measured for various load conditions and plotted as a function of the applied stress. Figures 3a shows the typical plot of normalized resistance change versus applied stress for p-type sensor. The responses of the sensors to applied uniaxial stress are highly linear. The difference of the resistance changes were extracted from the data in Fig. 3a and are plotted against stress in Fig. 3b.

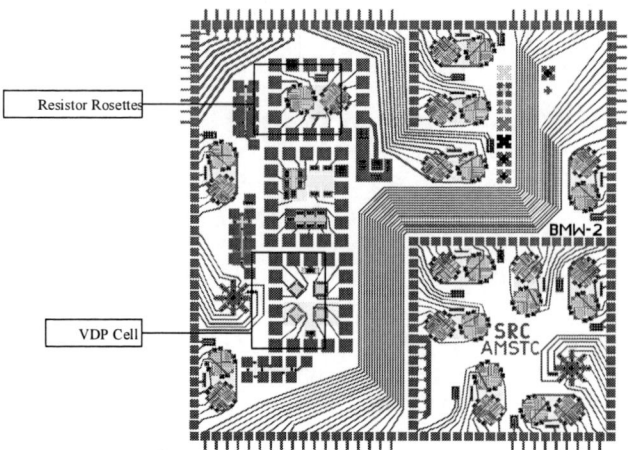

Figure 2 - (111) silicon test chip layout with VDP sensors

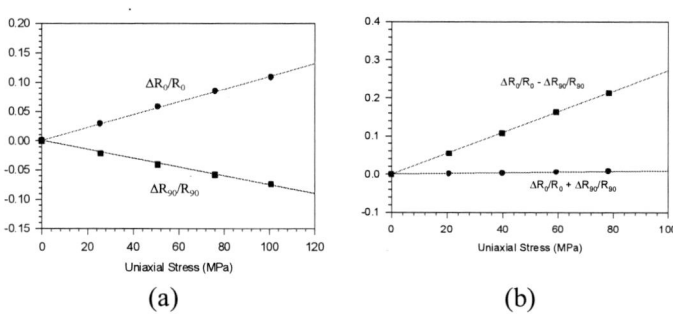

(a)   (b)

Figure 3 - Typical experimental data p-type sensor (17)

Similar tests were performed on several p- and n-type VDP sensors from the same wafer. The average values of the sensitivities are summarized in Table 1. Also given in Table 1 are the stress sensitivities for the analogous 0/90 resistor pairs, which were calculated using Eq. (6). The stress sensitivity in this case is $(B_1 - B_2)$ and were measured from the resistor rosettes. It is apparent form the tabulated experimental data that the responses of the difference of the normalized resistance changes to stress are 3.18 (1874.9/590.4) times for larger p-type and 3.32 (-413.9/-425.4) times larger for n-type sensors than the responses experienced by the analogous $0^{\circ}/90^{\circ}$ resistor sensor rosettes.

Sensor Type	Stress Sensitivity (1/TPa)
p-type VDP	+1874.8
n-type VDP	-1413.9
p-type Resistors	+590.4
n-type Resistors	-425.4

Table 1 – Stress Sensitivities for VDP and Resistor Sensors

IV. REFERENCES

[1] D. A. Bittle, J. C. Suhling, R. E. Beaty, R. C. Jaeger and R. W. Johnson, "Piezoresistive Stress Sensors for Structural Analysis of Electronic Packaging," *Journal of Electronic Packaging*, Vol. 113(3), pp. 203-215, 1991.

[2] J. C. Suhling and R. C. Jaeger, "Silicon piezoresistive stress sensors and their application in electronic packaging," *IEEE Sensors Journal*, vol. 1, no. 1, pp. 14-30, June 2001.

[3] A. Mian, J. C. Suhling and R. C. Jaeger, "The van der Pauw Stress Sensor," *IEEE Sensors Journal (in press)*.

Resistance Switching in $Sn_xMn_yTe_z$ –Based Devices

Patrick K. Herring[1] and Kristy A. Campbell[1,2] *Senior Member IEEE*

[1]*Micron Technology, Inc., 8000 S. Federal Way, Boise, ID 83706, USA;* [2]*Department of Electrical and Computer Engineering, Boise State University, 1910 University Dr., Boise, ID 83725*

Abstract – **Electronic devices comprised of a single layer of the chalcogenide material $Sn_xMn_yTe_z$ ($35 < x,z < 50$; $0 < y < 28$) deposited between two electrodes were fabricated and electrically tested for their ability to switch between high and low resistance states. Materials that have the ability to resistively switch when a potential is applied across them are promising as new electronic memory materials. In this work, the type of electrodes and the concentration of Mn present in the $Sn_xMn_yTe_z$ layer were both found to influence the ability of the device to resistively switch. When the top electrode material was Ag, resistance switching was observed. Additionally, only $Sn_xMn_yTe_z$ films with concentrations of Mn between 3 and 10 % were found to exhibit consistent switching.**

Keywords – memory, resistance variable, chalcogenide, switching.

I. INTRODUCTION

Chalcogenide glasses doped with Ag have recently been explored for use as electronic memories [1,2]. Typical chalcogenide materials used for these memories are Se-based glasses such as As_xSe_y or Ge_xSe_y doped with Ag. These electronic memories operate by application of a potential across the device that moves Ag^+ from an Ag-source layer at one of the electrodes into the Ag-doped glass, thus lowering the device resistance by forming a metallic-like conductive path between the two electrodes. The device resistance is returned to a higher resistance state by reversing the potential across the device electrodes, thus moving Ag^+ out of the chalcogenide glass back into the Ag-source layer. The electronic memory is thus defined by the two logic states: low resistance and high resistance.

Te-based chalcogenide glasses without Ag have been under investigation for many years for their use in phase change electronic memories [3,4]. Unlike the Ag-doped glasses, these glasses exhibit resistive switching by reversibly changing phase from amorphous to crystalline.

In this paper, we show that resistive switching occurs in a Te-based glass, $Sn_xMn_yTe_z$ with $3 < y < 10$ at. %, when an Ag-source electrode layer is present. Without this layer, this switching is not observed. Additionally, switching is not observed, even with an Ag-source layer, for high concentrations of Mn (> 20 at. %). Although switching can be observed in SnTe in the presence of an Ag-source layer, it is less consistent, and typically requires higher potentials, than the switching observed in films with low concentrations of Mn (3 to 10 at. %).

II. EXPERIMENTAL PROCEDURES

Thin films of $Sn_xMn_yTe_z$ were prepared by thermally co-evaporating SnTe (Alfa Aesar, 99.999% purity) and Mn (Alfa Aesar, 99.9% metals basis) using a CHA Industries SE-600-RAP thermal evaporator equipped with three 200 mm wafer planetary rotation. In order to obtain samples with a desired ratio of Mn to SnTe, the power to each source was manually adjusted to control the relative rates of deposition, creating on average the desired stoichiometry of the ternary material. Devices were created with films containing Mn concentrations in three ranges: (1) 0, (2) 3-10 at. % (low), and (3) 18-28 at. % (high). The Mn concentrations of the films tested varied significantly from experiment to experiment within these ranges due to the poor repeatability of the manual co-evaporation process. Film thicknesses were typically between 500 and 550 Å. Using the evaporator planetary rotator, films were deposited on three types of wafers simultaneously in each experiment: (1) a suprasil quartz wafer, (2) a Si wafer substrate with the layers 350Å W/800Å Si_3N_4 and, (3) a wafer processed for device fabrication. The two film characterization wafers present in each evaporation step were necessary in order to correlate the material properties to the device electrical characteristics since the manual co-evaporation process has poor repeatability.

Several analytical methods were used to characterize the $Sn_xMn_yTe_z$ films including Raman, UV-Vis, SEM, AFM, XRD, XPS, and ICP. However, here we will show only the XPS data and ICP data for SnMnTe film stoichiometry. XPS was collected with a Perkin Elmer PHI 5600 ESCA System PHI Quantum 2000; ICP, using a Varian Vista-PRO radial ICP. The SnMnTe films were removed from the wafer prior to ICP analysis with an etching solution of 1:1 $HCl:HNO_3$.

Electrical measurements were made using a Hewlett-Packard 4145B Parameter Analyzer. The fabricated devices were in 0.25 um diameter vias. The bottom electrodes were W and the top electrodes were either TiN or Ag.

III. RESULTS AND CONCLUSION

The XPS profiles through a low Mn concentration sample is shown in Fig. 1. For the low and high (not shown) Mn concentration samples the Mn concentration appears highest in the center of the film stacks. Additionally, each sample has a significant amount of oxygen present throughout the film.

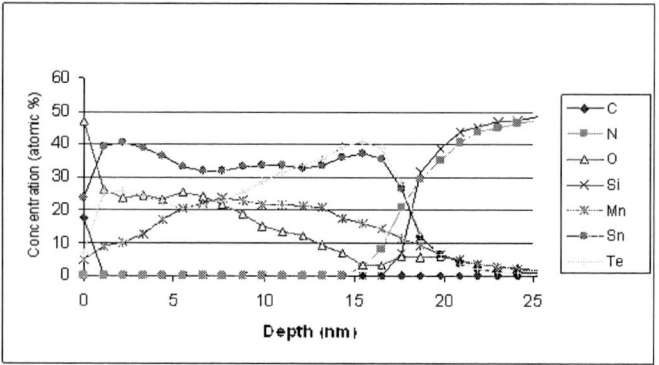

Fig. 1. XPS sputter profiles through the $Sn_xMn_yTe_z$ film with x = 43.2, z = 48.5, and y = 8.3 at. %. The x-axis scale is not an accurate representation of the film thickness.

A static IV trace for an SnTe device (no Mn) is shown in Fig. 2. Resistive switching is typically not observed at the low potentials shown in these samples. Switching between resistance states was observed only in the devices that had Ag top electrodes and low Mn concentration (Fig. 3). No resistance switching was observed in the high Mn concentration films (Fig. 4).

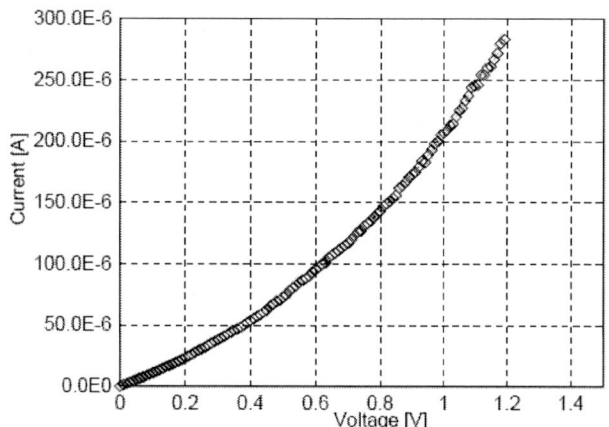

Fig. 2. Static IV trace of an SnTe device with an Ag top electrode. No resistance switching is observed in this case or in the case of a TiN top electrode.

The observed resistive switching in the low Mn concentration $Sn_xMn_yTe_z$ films only in the presence of an Ag electrode indicates that the switching is most likely due to Ag^+ incorporation into the $Sn_xMn_yTe_z$ material during application of a potential across the device.

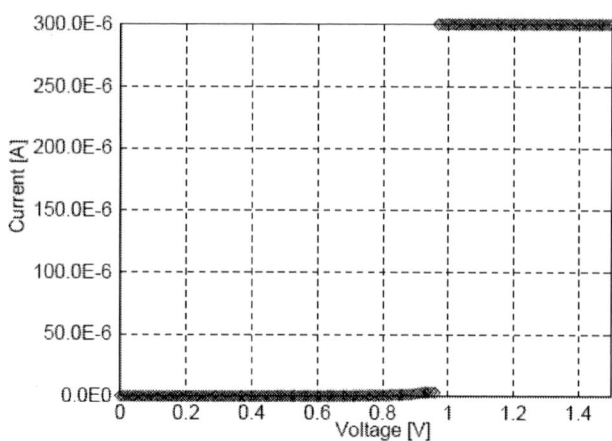

Fig. 3. Static IV trace of an $Sn_{47.9}Mn_{4.4}Te_{47.7}$ device with Ag top electrode. Notice the abrupt transition from non-conducting to conducting at about 0.95 V. This transition marks the resistance switching threshold from an OFF state to an ON state. No switching is observed in the case of a TiN top electrode.

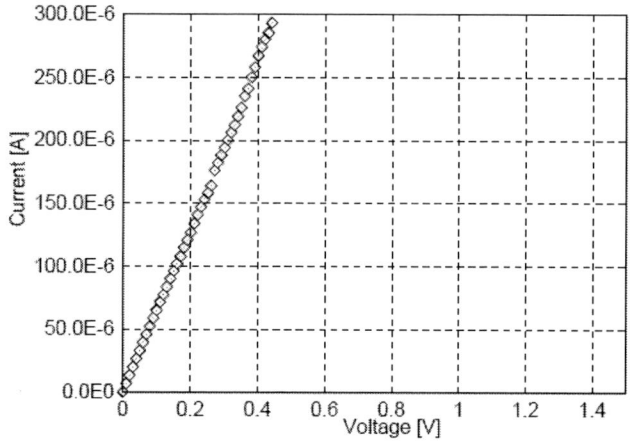

Fig. 4. Static IV trace of a $Sn_{40.6}Mn_{17.5}Te_{41.9}$ device with an Ag top electrode. No resistance switching is observed in this case or in the case of a TiN top electrode.

REFERENCES

1. Kund, M.; Beital, G.; Pinnow, C.-U.; Rohr, T.; Schumann, J.; Symanczyk, R.; Ufert, K.-D.; Muller, G. in IEDM, Tech. Dig. 2005, 31.5.
2. Mitkova, M.; Kozicki, M. N. Journal of Non-Crystalline Solids (2002), 299-302 (Pt. B), 1023-1027.
3. Neale, R. G. and Aseltine, J. A. *IEEE Transactions on Electron Devices* **20**, (1973) 195-209.
4. Lankhorst, M. H. R.; Ketelaars, B. W. S. M. M.; Wolters, R. A. M. *Nature* **4**, (2005) 347-352.

Metal/Semiconductor Contacts for Organic Molecules

Divesh Kapoor, Justin B. Jackson and Mark S. Miller, *Department of Electrical and Computer Engineering, University of Utah, Salt Lake City, Utah*

Abstract – **Contacts were developed to contact organic molecules to a metal or semiconductor substrate. Metal contacts were developed for thiol based chemical organic molecules and semiconductor contacts were developed for silane-based chemistry of molecules. The development of these contacts enables us to study electronic properties of organic molecules.**

I. INTRODUCTION

The current improvement in feature sizes of silicon-based technology will run into severe physical limitations as well as economic restrictions. Multiple prospective alternatives to complement or to replace this technology have been considered. One of them, molecular electronics involves the miniaturization by utilizing single molecules that would act as functional electronic components. The molecules are typically small organic molecules, which are about two orders of magnitude smaller than the present feature sizes. They appear to be the perfect components for designing future high-density electronic devices. Molecular electronics aims at individual contact to single molecules or small array of identical and perfectly ordered molecules.

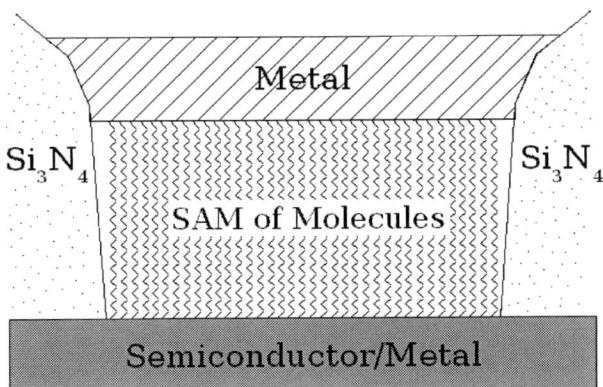

Figure 1: Metal/Semiconductor-Molecule-Metal Structure

The development of scanning probe techniques and progress in microfabrication processing allows the handling of single molecules or a monolayer of molecules[1] (SAM – Self Assembled Monolayer). However sophisticated and expensive equipment does not satisfy the basic requirement for molecular electronics that is to fabricate and package a functional hybrid integrated circuit. If we want to measure a current though individual molecule, we need a pair of electrodes spaced on the order of nanometers depending on the length of the molecules. If we are interested in measuring the electronic characteristics

of a monolayer of molecules, again we need contact electrodes, which must be controlled on the scale of nanometers length. The usual way is to utilize metal or semiconductor electrodes.

The best-investigated metal-based structure is the bond between a thiol (sulfur) group on the molecule and a gold substrate[2,3]. The thiol end group forms covalent bonds with the gold film. However, gold is not compatible with CMOS processing. Hence to study the electronic properties of single molecules or a monolayer of molecules with hybrid semiconductor molecular structures is fundamentally important[4].

II. PROCESS STEPS

For the silicon-molecule hybrid structure the process starts with a clean p-type 2" wafer with a thermally grown 700 nm silicon dioxide. The oxide is completely etched away to get a new clean silicon surface. A clean smooth silicon surface is very important for fabricating contacts for organic molecules. A field oxide 600 nm is grown using thermal wet oxidation. Using photolithography a set of alignment and focusing marks for e-beam lithography were etched into the field oxide. With the same mask a 50 μm window is created where the complete field oxide is etched away.

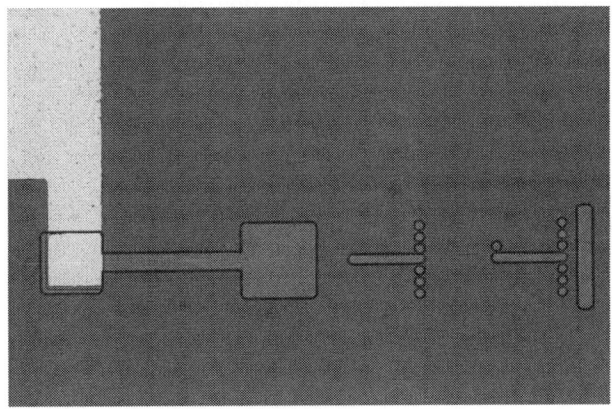

Figure 2: Area of Interest - Silicon Contact for Organic Molecules

Within this area of interest, the channel and contact region, phosphorus is diffused with expected sheet resistance of about 6 ohms/square. A 25 nm thin oxide is grown by means of dry oxidation. On top of this thin oxide a 35 nm thin layer of silicon nitride is deposited. The silicon nitride is deposited using low-pressure chemical vapor deposition (LPCVD) at 825 °C by reacting 60 sccm dichlorosilane and 10 sccm of ammonia. Using higher ratios of DCS to ammonia results in a lower stress film. The thin silicon dioxide layer below the silicon nitride layer acts as an etch stop in dry etching of silicon nitride. It is followed by photolithography with mask 2 and mask 3 to create contact vias and aluminum contact pads for the n-type phosphorus doped layer and for the substrate. The wafer is then annealed at 475 °C for 20 minutes, to get ohmic contacts. The wafer was diced into individual die.

For e-beam lithography, a 100 nm thin layer of e-beam resist PMMA 950K is spun on and baked on a hot plate at 170 °C for 15 minutes. A 50 nm square pattern is exposed into the PMMA, with a line dose of 0.6 µC/cm. The exposed PMMA is developed with MIBK:IPA for 30 seconds. This exposes the underlying bilayer of silicon nitride and silicon dioxide. The silicon nitride is etched using RIE etching and the silicon dioxide is etched with BOE, which exposes the doped silicon layer. The PMMA is removed with dichloroethane, which strips the PMMA much better than acetone[5].

For a metal-molecule structure, the nanopore concept is utilized. The process starts very similar to the silicon-molecule structure. Once the alignment, focusing marks and the region of interest, a 50 µm x 50 µm square, have been lithographically etched into the wafer, a 45 nm low-stress Si_3N_4 film is deposited by LPCVD on both sides of the wafer. A 450 µm x 450 µm window is etched into the backside of the wafer dies, aligned to the area of interest on the front side of the wafer. By means of e-beam lithography, a single hole of approximately 30-50nm is etched into the Si_3N_4 film on the front side.

Figure 3: Silicon Nitride Membrane with a Nanopore

An anisotropic etch, a 30% KOH solution at 85 °C is used to etch the silicon, from the back side, to obtain a silicon nitride membrane with a pore. The bottom Au electrode is then placed by sputtering 200 nm of gold to fill the pore.

III. RESULTS

These two processes can be used to create metal/semiconductor electrode in the nanometer scale. On these electrodes, a self-assembled monolayer of molecules of interest can be formed with a metal top contact to make electrical measurements.

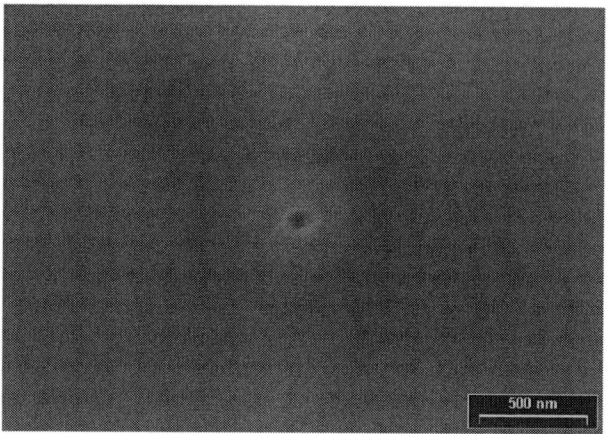

Figure 4: A 35 nm diameter pore in Silicon Nitride film

If needed, smaller pores can be fabricated by using e-beam lithography setup at other facilities. The advantage silicon-molecule structure is that it can be easily be integrated with CMOS process steps, in case some interesting useful electrical characteristics are observed.

IV. REFERENCES

[1] M. C. Hersam, N. P. Guisinger, J. W. Lyding, "Silicon-based Molecular Nanotechnology" *Nanotechnology*, Vol 11, pp. 70–76, 2000.

[2] K. Kobayashi and T. Fukuma, "Alkanethiol self-assembled monolayers on Au(111) surfaces investigated by non-contact AFM " *Applied Physics A: Materials Science and Processing*, Vol. 72, pp. 109-112, 2001.

[3] Takhee Lee Wenyong Wang and Mark A. Reed, "Mechanism of electron conduction in self-assembled alkanethiol monolayer devices", *American Physical Society: Physical Review B*, Vol. 68, pp. 035416-1 to 035416-7, 2003.

[4] Christopher E. D. Chidsey and Mathew R. Linford, "Alkyl monolayers covalently bonded to silicon surfaces", *Journal of American Chemical Society*, Vol. 115, pp. 12631-12632, 1993.

[5] Qingling Hang, Davide A. Hill, Gary H. Bernstein, "Efficient removers for polymethylmethacrylate (PMMA)", *Journal of Vacuum Science and Technology B*, Vol. 21 , No.1, pp. 91-97, 2003.

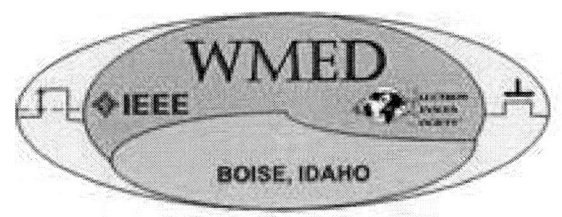

Session

Processing and Reliability

Space Efficient ESD Methodology for Reliable High Volt Applications

John J. Naughton, Matthew Tyler, Muhammad Anser
AMI Semiconductor
2300 Buckskin Rd, Pocatello, Id 83201
208-234-6062; fax: 208-234-6740; e-mail: John_J_Naughton@amis.com

Abstract— **LDMOS-SCR structures have been shown to provide formidable ESD protection. This work characterizes CMOS process regions widely used in high voltage technologies to control triggering characteristics >40V. This is done using the breakdown voltage of a standard gate poly structure and structures without gate oxide further enhancing reliability during an ESD event.**

Keywords-component; ESD, High Voltage, SCR, Relaibility

I. INTRODUCTION

Analog mixed signal devices for high reliability applications such as automotive and implantable medical devices present a variety of design challenges. These challenges include interleaved high and low voltage domains, bulk silicon isolation issues, limited pin count, and die size restrictions. Utilizing standard CMOS processing with minimal additional steps provide a low cost, high reliability method of high volt device manufacturing [1]. Variations of standard n-channel field dopant concentrations and diffusion to well layout rules can be used to build LDMOS devices in a single well technology with up to 40 V operation. Careful tailoring of dopant profiles is utilized to define extended MOS drains and control their subsequent breakdown voltages (BV).

In the current work, a method of targeting ESD protection threshold (Vt1) is demonstrated that provides greater reliability and space efficiency in a standard CMOS, single well technology for high volt applications. This is accomplished by constructing a Silicon Controlled rectifier (SCR) protection device around a standard I/O pad.

II. METHODOLOGY

In this methodology, the BV of a known device is used to set the trigger voltage (Vt1) of an SCR. SCR type structures are an established method of ESD protection for CMOS, BiCMOS, multiple-supply, mixed-signal, parasitic-sensitive RF and high-pin-count ICs [2,4,5]. Once triggered, the SCR provides a low impedance path to ground protecting both the core circuitry and base silicon substrate. The pad-oriented structure can accommodate customer applications for minimal die size and pin count [5]. This configuration creates a large bi-polar area within a small die size area that would normally be consumed only by the bond pad. By virtue of its placement, the pad oriented SCR can more efficiently contain the ESD energy at the injection site (bond pad) [5]. The base layer implants are constructed symmetric with respect to the X and Y-axis of the pad to offset any mask alignment error that may

be incurred. This symmetry significantly reduces the die-to-die variation of Vt1.

Another challenge is that spacing rules and dopant profiles tend to have non-linear characteristics. Key to controlling breakdown voltages in a nearly linear fashion is to optimize the Nwell boundary profile (n-channel field enclosure of n-tub) for a breakdown that is approximately equal to 1.5*Vdsmax. A high concentration implant of N+ diffusion near the Nwell edge will contribute carriers near the depletion region, which can be quite large in a low concentration Nwell (often needed for high voltage applications). Traditionally, this interaction is avoided due to a strong dependence on mask alignment. As a result, designers will modulate the implant density to control avalanche breakdown. The inherent symmetry of the pad oriented protection structure works to average mask alignment errors allowing the designer to use this progressive junction control technique with reduced risk. Once characterized, the n-tub enclosure of N+ active can be dialed in to optimize each individual application to maximize the ESD window.

III. EXPERIMENTAL

In this example, the SCR ESD protection structure is constructed in a 0.5um 20V technology. The structure is triggered by the avalanche breakdown of the N-well region. The Vt1 of the protection structure should then be very close to the BVDSS of an NLDMOS for this technology. The drain of the NLDMOS was left floating, the emitter of the PNP was connected to pad metal, but a soft connection was included (N+ in N-well) from pad metal to the drain of the NDLMOS interior to the PNP emitter. This soft connection forces Nwell current to flow through the base of the vertical PNP (Figure 2).

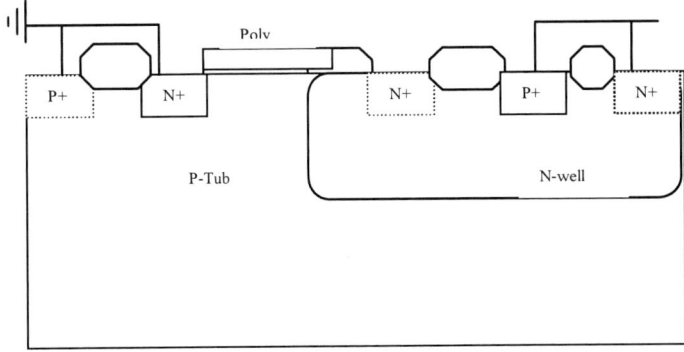

Figure 1.   Cross Section of the Nldmos triggered scr esd protection structure.

Figure 2 illustrates the transmission line pulse (TLP) results, a Vt1 of ~27V, a holding current >100mA, and a device damage threshold (Vt2) > 10A. This proved ideal with a Vt1 greater than normal operating voltage (20V) and less than the VDSS of the core circuitry (29V).

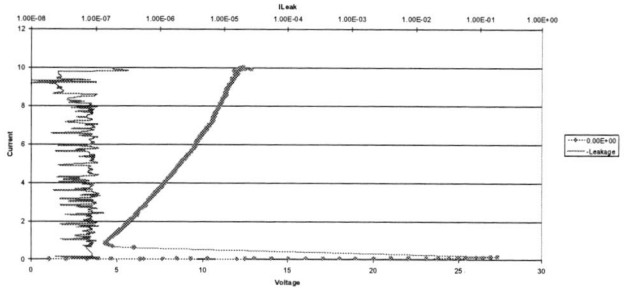

Figure 2. TLP results from the Nldmos triggered scr esd protection structure.

Eliminating gate oxide would provide a more reliable ESD protection structure. The potential for charge trapping near the oxide edge can reduce the Vt1 with each successive pulse [4]. Figure 3 shows a cross section of a structure using a traditional SCR design within a 20V well but transposing the N+ and P+ implants to encourage the proper base current flow. It alsp illustrates the 0.5 um enclosure of the N-well with n-channel field, a layout rule for 20V well isolation in this technology.

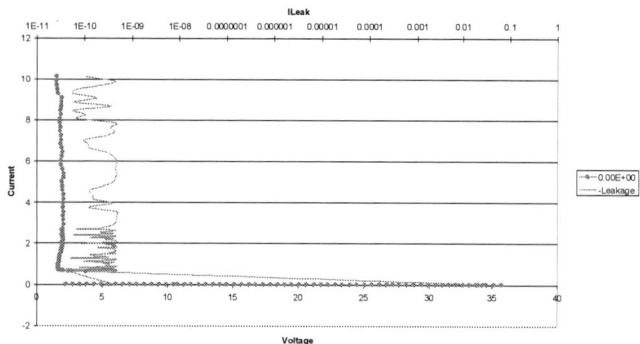

Figure 3. Modified Nwell triggered scr esd protection structure

The Vt1 of this structure proved much higher than suitable for a 20V technology, 37V (Figure 4). It did show good snapback, good holding current (100mA) and a Vt2 greater than 10A.

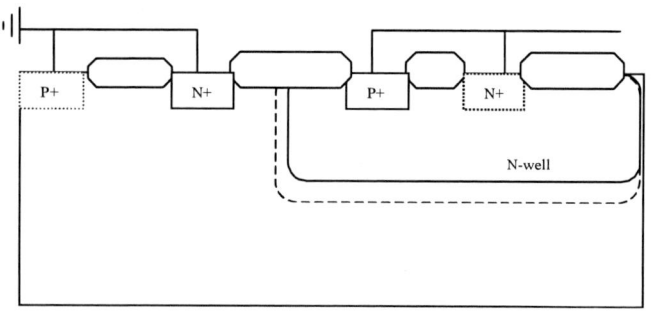

Figure 4. TLP results from the revised scr design-no gate poly

This structure did prove useful in custom circuit protection outside the standard voltage domains such as short to battery protection and I/O voltage tolerances.

TCAD simulations have shown that by varying the overlay of implants in a given technology, in this case the n-channel field, and/or the proximity of high dopant concentration diffusion implant near the well boundary, you can accurately target the Vt1 for the given circuit need without additional masking steps. For example, with a 1um minimum N+ spacing to N-well edge in the N-well and a 1um minimum spacing N+ to N-well edge in the P-tub, a Vt1 of 30V is achieved. If the N+ to N-well spacing is reduced to 0.5um, then the N-well breakdown is ~20V. This layout technique provides a range of Vt1 of the SCR from 20-30V, optimizing the Vt1 for the maximum "ESD window". Figure 5 shows the results of these simulations.

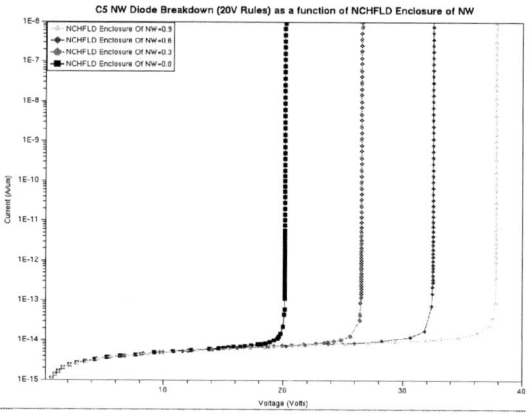

Figure 5. T-Cad simulations showing VT1 as a function of overlay

IV. CONCLUSIONS

A space efficient High Volt SCR ESD protection structure has been proven by utilizing pad orientation. Two types of structures have been incorporated into this methodology; an NLDMOS triggered structure for precise Vt1 targeting, and a high reliability structure by eliminating gate oxide that is triggered by a parasitic bipolar. By varying the n-channel field enclosure of the N-well, the Vt1 of the protection structure can be customized for the device application. T-cad simulations for N-well to p-sub breakdown can be used to determine this Vt1 prior to actual silicon.

V. REFERENCES

[1] G.M. Dolny, et al, "Enhanced CMOS for Analog-Digital Power IC Applications", *IEEE Trans. Elec. Dev., 1986* p 1985

[2] V.A. Vashchenko and M. ter Beek, ESD Protection Window Targeting Using LDMOS-SCR Devices with PWELL-NWELL SUPER-JUNCTION". *IRPS 2005* pp 612-613

[3] S. Xu et al, "120V Interdigitated-Drain LDMOS (IDLDMOS) on SOI Substrate Breaking Power LDMOS Limits," *IEEE Trans. On Elec. Dev.,* 47, N10, 2000, pp. 1980-1985

[4] Bart Keppens, et al, "ESD Protection Solutions for High Voltage Technologies", Sarnoff, Belgium

[5] Bonding-Pad-Oriented On-Chip ESD Protection Structures for IC's, H. Feng et al., ISCAS 2003, Vol 1, 25-28, May 2003, pp. I-741-I-744.

Time-Dependent Dielectric Breakdown of a Recessed Channel DRAM Access Device

T. Owens, D. Hwang, P. Vaidyanathan, and K. Parekh

R&D Process Development, Micron Technology, Inc. Boise, ID

(Phone) 208 368 2846, (Fax) 208 363 2919, (e-mail) tjowens@micron.com

Abstract—A recessed access device (RAD) used in a DRAM cell has exhibited advantages over the conventional planar access device, including retention time improvement. However, worse Time-dependent dielectric breakdown (TDDB) characteristics were observed for RAD. The degraded TDDB performance is primarily attributed to thinner oxide growth in the recess.

Index Terms—TDDB, oxide reliability, power-law, access device, recessed access device (RAD), DRAM.

I. INTRODUCTION

As DRAM technology advances to meet the demand for increased bit density, scaling of the Access Device (AccDev) becomes a significant challenge. This challenge has prompted the evaluation of three-dimensional (3D) structures for the AccDev, which offer advantages in leakage characteristics and scalability [1-2]. In addition to meeting retention time and transistor drive requirements, these 3D devices must be robust from a reliability perspective. This paper presents the findings from a time-dependent dielectric breakdown (TDDB) study of a recessed access device (RAD) and compares the results with those from a conventional access device.

II. EXPERIMENTAL

RAD structures were fabricated by integrating a trench etch into a baseline DRAM process flow. Fig. 1 shows a TEM of the RAD structure.

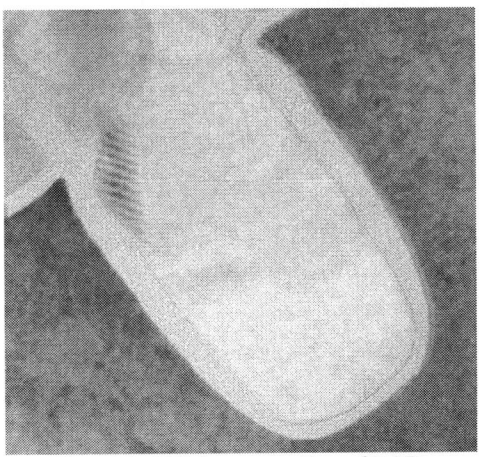

Figure 1: TEM cross-section of RAD.

Amorphous Si (a-Si) is used to create the gate electrode. Phosphorus or Boron implantation created n-type or p-type gate electrodes. Both npoly and ppoly RAD were compared with ppoly planar devices.

TDDB characterization was performed by constant voltage stress (CVS) at several gate voltages. Time-to-failure (TTF) was recorded at the first breakdown event. Breakdown detection consisted of monitoring stress (I_{STR}) and use-condition (I_{USE}) current for a discontinuous increase in current. An increase of 100X in I_{STR} or I_{USE} was classified as oxide breakdown. All CVS measurements were performed under inversion mode stress at a temperature of 100°C. The test structures consisted of 120-device and 300,000-device arrays of individual access devices, equivalent to gate oxide areas in the range of 1 to 3,000μm².

Weibull statistics were applied to the oxide breakdown data. The characteristic lifetime (T63) and shape factor (β) were determined from the breakdown distributions. These parameters were used in extrapolating TDDB lifetime to the operating condition.

III. RESULTS AND DISCUSSION

Fig. 2 shows retention-time characteristics of RAD and planar access devices. RAD shows a substantial improvement over the planar structure, which makes it a desirable replacement for the conventional (planar) DRAM access device.

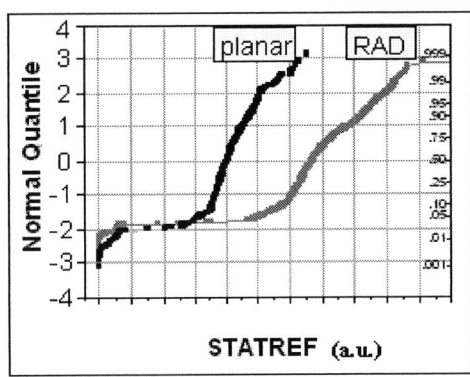

Figure 2: Retention-time characteristics.

Fig. 3 shows TTF vs. Vg results for a ppoly planar access device, ppoly RAD, and npoly RAD. These results demonstrate that the RAD structure has shorter TTF than the planar structure. At Vg = 5.5V, the npoly RAD TTF is approximately 160X shorter than ppoly . While, for ppoly RAD, the TTF ratio is about 35. The TTF disparity between npoly and ppoly RAD is due to the workfunction difference of the gate electrodes, which is ~1V.

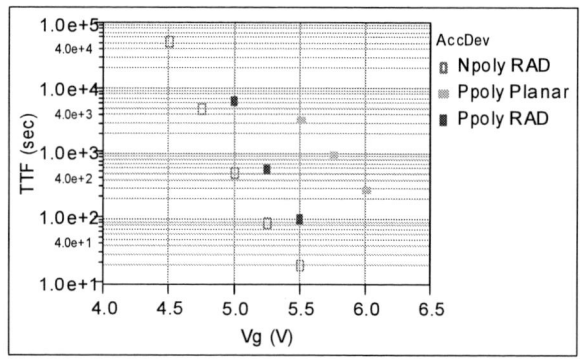

Figure 3: Time-to-fail (T63%) vs. Vg.

The larger workfunction for ppoly results in a smaller voltage drop across the oxide (Vox) for a given applied Vg. This is evident in Fig. 4, which shows median gate leakage (Igate) at each Vg stress condition for the different access device structures. It is also clear from Fig. 4 that the ppoly RAD has about 50X more leakage than the ppoly planar device at Vg = 5.5V. Based on geometrical considerations and capacitance measurements, the RAD area is at most 2X larger than the planar device. Since the tunneling current is directly proportional to area, the leakage current increase due to area should only be 2X. This discrepancy between expected and observed leakage currents is attributed primarily to a difference in oxide thickness (Tox) between the RAD and planar structures. Indeed, TEM images (see Fig. 1) indicate non-uniform gate oxide growth, with a thinner oxide formed at the bottom corners of the recess.

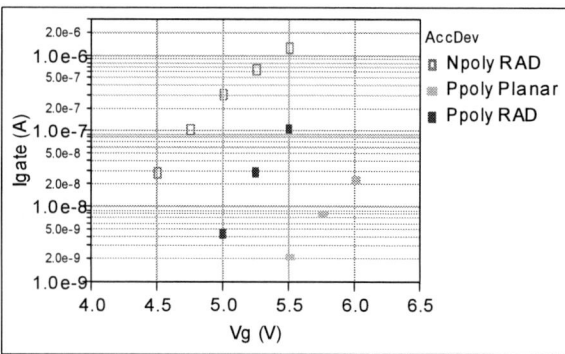

Figure 4: Median value of Igate vs. Vg.

The reduced TDDB lifetime for RAD is understandable in light of the thinner Tox and increased Igate. It is noted that even for equivalent oxide thickness, RAD has a larger oxide electric field (Eox) than planar due to the curvature of the recess. The increased Eox results in larger Igate, and therefore reduces TTF as well. The area difference also affects TTF based on the area scaling relationship, Eq (1).

$$TTF1/TTF2 = (A2/A1)^{(1/\beta)} \qquad (1)$$

For an area ratio of 2, with β = 2.5, the TTF ratio is approximately 1.3, which is much smaller than the observed TTF ratio. This result, combined with the observation that charge-to-breakdown (Qbd) is quite similar for RAD and planar devices, supports the conclusion that the TTF difference is principally a result of thinner oxide and the associated fluence increase. A similar conclusion of reduced oxide thickness was reached by other authors [3].

The npoly RAD TTF vs. Vg results are re-plotted in Fig. 5 on a log-log scale. The TTF has been scaled to a 2Gb DRAM and 0.01% failure level (100ppm). Data for Vg ≥ ~5V fits a power-law (TTF ~ V^{-n}), with a value of ~33 for the exponent, n. However, for Vg < ~5V, the power-law exponent is ~44. Similar results were presented recently in [4]. The slope change can be explained by a sharp decrease in defect generation efficiency below a Vg of ~5V. As evident from Fig. 5, the exponent change has a significant impact on lifetime projections.

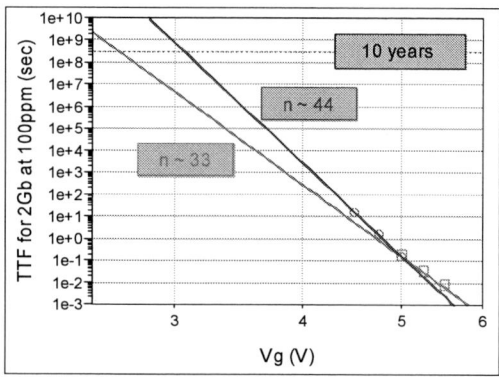

Figure 5: TTF for 2Gb DRAM at 100ppm failure level. Below Vg = ~5V, the power-law slope increases from ~33 to ~44.

IV. CONCLUSION

While the degraded TDDB performance of RAD is a reliability concern, it does not exclude RAD as a viable alternative to the conventional DRAM access device. Experimental results indicate that the primary cause of TDDB lifetime reduction is thinner oxide in the recess. Therefore, gate oxidation process optimization may be able to recover much of the lost TDDB margin. Also, TDDB data collected at lower Vg agree with a recently proposed voltage acceleration model that predicts a more optimistic lifetime, indicating that present TDDB requirements may be too strict.

REFERENCES

[1] C.J. Radens *et al.*, "An Orthogonal 6F2 Trench-Sidewall Vertical Device Cell for 4Gb/16Gb DRAM," IEDM 2000.
[2] J.Y. Kin *et al.*, "The Breakthrough in data retention time of DRAM using Recess-Channel Array Transistor (RCAT) for 88nm feature size and beyond," VLSI Symp 2003.
[3] J.Y. Seo *et al.*, "Reliability for Recessed Channel Structure n-MOSFET," *Microelectronics Reliability*, vol. 45, pp. 1317-1320, 2005.
[4] R. Duschl *et al.*, "Is the power-law model applicable beyond the direct tunneling regime?" *Microelectronics Reliability*, vol. 45, pp. 1861-1867, 2005.

Preliminary Study of NOR Digital Response to Single pMOSFET Dielectric Degradation

T. L. Gorseth[1], D. Estrada[1], J. Kiepert[1], M. L. Ogas[1], B. J. Cheek[1],
P.M. Price[2], R. J. Baker[1], G. Bersuker[3], W. B. Knowlton[1, 2]

[1]*Dept. of Electrical and Computer Engineering,* [2]*Dept. of Materials Science and Engineering, Boise State*
University, Boise, ID 83725 USA, [3]*SEMATECH, Austin, TX 78741 USA*

Abstract—**The voltage-time domain (VT) characteristics of the CMOS NOR logic circuit are investigated using a switch matrix technique (SMT). VT performance is analyzed following gate oxide wearout in a pMOSFET, induced by applying a constant voltage stress (CVS) at -4.0 V. Results for the NOR VT characteristics show approximately 30% increase in rise time (t_r), a considerable digression from nominal operation.**

I. INTRODUCTION

THE effects of gate dielectric breakdown (BD) on digital circuit operation is becoming progressively more important as gate oxide thicknesses (t_{ox}) are scaled below 2.0 nm [1], due in part to increased gate leakage current. Therefore, better understanding of BD and its effect on circuit function may lead to more accurate reliability projections [2]. Through simulation performed on digital circuits, such as NAND and NOR gates, it has been reported that BD leads to considerable timing delays [1], [3]. However, the physical characteristics pertaining to the effects of oxide degradation in both the devices and circuits have received less attention.

A recent study by Ogas *et al.* examined the effects of single degraded pMOSFET devices on the voltage-time response (VT) of the NAND logic circuit [4]. Using a SMT developed by several of the co-authors [5]-[6], they were able to conclude that a substantial increase in rise time (t_r) was attributed to an increase in channel resistance of the degraded pMOSFET [4].

In this preliminary study, the same wearout degradation regime and circuit reliability technique reported by Ogas *et al.* [4] is targeted to examine the VT response of the NOR logic circuit.

II. EXPERIMENTAL

The MOSFET devices used in this study are fabricated using a 0.1 μm CMOS process with a t_{ox} of 2.0 nm. The device width is 10 μm and the length is 0.1 μm. Measurements are obtained using an integrated semiconductor characterization system described elsewhere [6]. SMT involves wafer level measurements using two probe stations, each with eight micropositioners necessary to connect the NOR circuit (Fig. 1). This technique permits the measurement of NOR circuit performance, as well as, the isolation of four

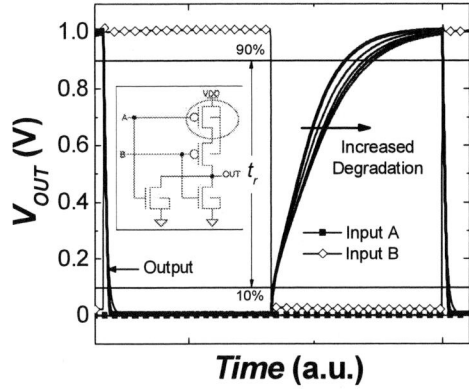

Fig. 1. NOR Voltage-Time measurement with Input A held at ground and Input B swept from 0V to 1V (configuration 2). Inset: NOR circuit with degraded pMOSFET circled.

separate MOSFET devices to obtain individual MOSFET characteristics [6].

Gate oxide wearout in one pMOSFET of the NOR gate is induced by applying constant voltage stress (CVS) at -4.0 V in periods of 600 seconds (Fig. 2). This approach has been used in similar studies involving pMOSFET devices [7]. The remaining three MOSFETs in the NOR circuit are not stressed. Wearout is the cumulative effects of oxide degradation over five CVS periods, as defined by Ogas *et al.* [4]. After each stress period, NOR circuit VT and device characteristics were measured which include gate leakage current (I_G-V_G) and maximum drain current ($I_{DRIVE,MAX}$).

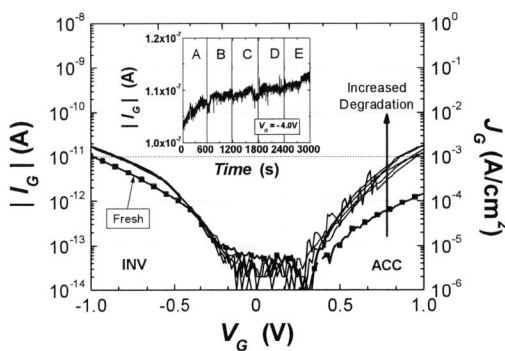

Fig. 2. Pre- and post-CVS I_G–V_G data showing the effects of wearout in a pMOSFET. Inset: I_G vs. time, in intervals of 600s, during CVS.

TABLE I
NOR GATE INPUT CONFIGURATIONS INDICATING I/O STATE

Configuration	Input/Output			
	1	2	3	4
Input A	VDD	GND	Pulse	Pulse
Input B	Pulse	Pulse	VDD	GND
Output	0	0,1	0	0,1

Grey = positions affected by wearout. 0,1 = output transition

Table I illustrates the NOR gate response examined for the pMOSFET position highlighted in Fig. 1. It is observed that the circuit response is affected by only configurations 2 and 4. For this preliminary study, only configuration 2 is presented. The data for configuration 4 is currently being analyzed. However, initial measurements suggest configuration 4 yields similar results to configuration 2.

III. RESULTS AND DISCUSSION

NOR Circuit: The VT characteristics for the NOR circuit in configuration 2 are shown in Fig. 1. It was observed that t_r of the circuit increases by 27 % ±5 % for configuration 2. Rise time is defined as the time it takes for the output signal to increase from 10% to 90% of the final voltage, as indicated in Fig. 1 [8]. It should be noted that previous studies on inverter and NAND logic circuits have indicated that changes in t_r are indicative of pMOSFET degradation, while changes in fall time (t_f) are associated with the nMOSFET [4]-[6]. Moreover, Carter *et al.* suggests that an increased t_r could potentially cause logic errors in a high speed digital circuit when introduced into a large scale design.

MOSFETs: The increased t_r of the NOR circuit response warrants investigation of the characteristics of the degraded pMOSFET. The CVS graph coupled with the gate leakage (Fig. 2) suggests the pMOSFET has not endured a traditional dielectric breakdown (i.e. hard breakdown) [9]. Moreover, as indicated by Ogas *et al.*, the pMOSFET device appears to be operating in the wearout regime [4]. Further analysis of the data indicates a dramatic change in the performance of the pMOSFET device attributed to dielectric wearout (Fig. 3), despite the minimal change in gate leakage current (Fig. 2). Additionally, the $I_{DRIVE,MAX}$ is decreased by 41.0± 5.4 %, (Fig. 3).

Similar pMOSFET degradation was observed by Ogas *et al.* in [4]. They found the degradation produced an increase of t_r

in NAND gates. The t_r increase was correlated to an increase in channel resistance by relating the degradation in pMOSFET parameters to channel resistance. The results of Ogas *et al.* suggests the increase in t_r of the NOR circuit as observed in this study correlates to an increase in channel resistance of the degraded pMOSFET. Particularly interesting is the observation that t_r degradation from the NOR VT characteristics is approximately half that of the t_r degradation as compared to NAND circuit operation [4]. The difference in VT response may be due to the stacking nature of the pMOSFET devices in the NOR circuit compared to the pMOSFET configuration in the NAND circuit, justifying further examination [10].

IV. CONCLUSION

The preliminary results reported for wearout in one pMOSFET of a NOR gate circuit indicate a substantial increase in t_r attributed to an increase in channel resistance. This preliminary study suggests that the NOR circuit configuration is less sensitive to single pMOSFET wearout than the NAND circuit, potentially due to the stacking effect in digital circuits. Future goals include investigation of the stacking effect in digital circuits relative to the wearout regime, as well as examination of correlations between increased t_r and variations in threshold voltage.

ACKNOWLEDGMENT

Funding for the project was supported in part by the DoD Multidisciplinary University Research Initiative (MURI) program #F49200110374, NASA Idaho Space Grant, NASA Idaho EPSCoR, Idaho NSF EPSCoR Award #EPS-0132626, NSF MRI Award #0216312, DARPA Contract #N66001-01-C-80345, and NIH INBRE #P20RR16454.

REFERENCES

[1] A. Avellán and W. Krautschneider, "Impact of soft and hard breakdown on analog and digital circuits," *IEEE TDMR*, vol. 4, pp. 676-680, Dec. 2004.

[2] J. H. Stathis, "Impact of ultra thin oxide breakdown on circuits," in *2005 Proc. IEEE ICICDT*, pp. 123-127.

[3] J. R. Carter, S. Ozev, and D. J. Sorin, "Circuit-level modeling for concurrent testing of operational defects due to gate oxide breakdown," in *2005 Proc. IEEE DAT in Europe*, pp.50-55.

[4] M. L. Ogas, *et al.*, "Degradation of risetime in NAND gates using 2nm gate dielectrics," in *2005 Proc. IEEE IIRW.*, pp. 63-66.

[5] N. Stutzke, *et al.*, "Effects of circuit-level stress on inverter performance and MOSFET characteristics," in *2003 Proc. IEEE IIRW*, pp. 71-79.

[6] B. Cheek, *et al.*, "Investigation of circuit-level oxide degradation and its effect on CMOS inverter operation and MOSFET characteristics," in *2004 Proc. IEEE IRPS*, pp. 110-116.

[7] M. L. Ogas, *et al.*, "Survey of oxide degradation in inverter circuits using 2.0 nm MOS devices," in *2004 Proc. IEEE IIRW*, pp. 32-36.

[8] R. J. Baker, *CMOS: Circuit design, layout, and simulation*, 2 ed: IEEE Press Wiley-Interscience, 2005, pp. 50.

[9] S. Lombardo, *et al.*, "Dielectric breakdown mechanisms in gate oxides," *JAP*, vol. 98, pp. 1-36, Dec. 2005.

[10] S. Mukhopadhyay, A. Raychowdhury, and K. Roy, "Accurate estimation of total leakage current in scaled CMOS logic circuits based on compact current modeling," in *2003 Proc. Design Automation Conf.*, pp. 169-174.

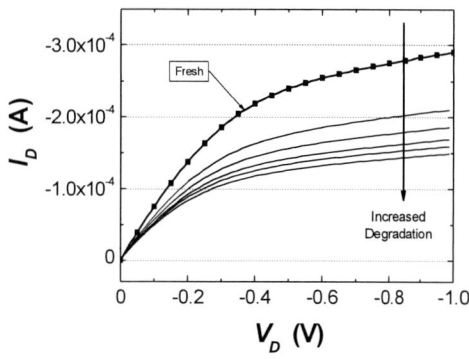

Fig. 3. Drain current vs. drain voltage characteristics of degraded pMOSFET from Fresh to Test E.

Ultra-Low-Power Dynamic-Threshold Digital Circuits
in the FlexFET Independently-Double-Gated SOI CMOS Technology

S. Parke+, K. DeGregorio*, D. Hackler*, D. Wilson*

*American Semiconductor, Inc., Boise, ID

+Boise State University, Boise, ID

Introduction

The ITRS roadmap projects that double-gated transistors will be needed in the future, in order to scale CMOS to the 45nm node. However, flexible, dynamic threshold control is possible now with existing independently-double-gated (IDG) CMOS technologies, and is highly desirable for ultra-low-power (ULP) SoC designs at the 180nm, 130nm, and 90nm nodes [1]. By varying the bottom gate voltage of the FlexFET IDG MOSFET from -0.5V to +0.5V, standby power can be dynamically changed over **ten orders** of magnitude, while the transistor/circuit performance can be changed by 70%. Minimally sized transistors may be used to achieve ULP in standby, while dynamic Vt adjustment is used to achieve high-performance when active. This paper demonstrates IDG FlexFET CMOS in static CMOS ring oscillators, while the advantages of applying IDG-CMOS to dynamic domino CMOS logic circuits have recently been shown as well [1]. IDG-CMOS has also recently been applied to several exciting new analog/MS/RF circuit applications, such as a single transistor mixer [5]

Device Description

FlexFET is a new SOI IDG-CMOS technology [2-4] that features a gate trench etched through thick SD regions into which an implanted JFET bottom gate and mid-gap TiN MOS topgate are self-aligned, as shown in Fig. 1. The independent top and bottom gates are contacted at opposite sides of the channel by a damascene tungsten local interconnect (LI) that is embedded in the isolation region between devices. This results in compact, planar connections to all four transistor terminals. Individual transistors can be connected as single gate (SG), double gate (DG), or (IDG) as required. Fig. 2 shows an SEM topview of an IDG inverter ring oscillator (taken after LI mask). All NMOS bottomgates are connected to NBG and all PMOS bottomgates are connected to PBG. Fig. 4 shows the schematics of the three inverter types investigated. The SG and IDG types can both operate to VDD=2.5V, while the DG type can only operate to VDD=1.0V. Fig. 3 shows that a measured bottomgate threshold control of >0.85V/V was achieved. This high degree of control means that only a small bottomgate voltage range of -0.5V to +0.5V is required to achieve the desired ultra-low-power and high performance targets. No additional area was required to make the bottomgate contacts.

Process Description

The FlexFET process uses a relatively thick 200nm SIMOX SOI layer into which SD's are implanted. Gate trenches are then etched through the SD's. The channel thickness is set by the energy of the bottomgate implant into the bottom of the gate trench, not by the SOI thickness. A raised SD structure is achieved without any epi growth thermal budget. FlexFET is a "gate-last" process, allowing high-K or ferroelectric gate dielectrics and metal topgates to be deposited and planarized into the gate trench, thus avoiding any damaging plasma gate etches or subsequent high-temperature processing. Ultrashallow SDE's are formed by disposable doped glass sidewall spacers. PECVD PSG & BSG conformal spacers with 5% dopant concentration were created on the gate trench sidewalls, used to drive in the SDE's, and then removed and replaced by nitride spacers. These nitride spacers permit 180nm channels to be created with relaxed 350nm lithography. A 4.5nm oxynitride and a 30nm TiN layer form the topgate stack. Only 8 masks were used through M1. The recessed channel & bottomgate structure maintains an ultra-thin 20nm channel while allowing 200nm thick SD's for lower SD series resistance. These unique device structural features make the FlexFET transistor highly scalable.

Results & Discussion

Fig. 5 shows the IDG-CMOS propagation delay as a function of both NBG and PBG. The slowest condition (44ps) is for both bottomgates "off" at -0.5V, while the fastest condition (26ps) is for both bottomgates "on" at +0.5V. The table in Fig. 6 shows that the threshold voltage can be varied by 0.85V, resulting in the standby current reduced to 10^{-16}A/um at -0.5V, while the drive current is increased to 850uA/um at +0.5V. Essentially all static & dynamic digital circuits, as well as analog/MS/RF circuits can benefit from IDG operation.

References

[1] H. Mahmoodi, et. al., 2004 IEEE SOI Conf., pp. 67-68, Oct. 2004.

[2] S. Parke, et. al., 2004 IEEE SOI Conf., pp. 104-105, Oct. 2004.

[3] S. Parke, et. al., 2005 IEEE WMED, pp. 35-37, April 2005.

[4] S. Parke, et. al., 2005 IEEE Aerospace Conf., 7.0902, Mar. 2005.

[5] L. Mathew, et. al., 2004 IEEE SOI Conf., pp. 187-189, Oct. 2004.

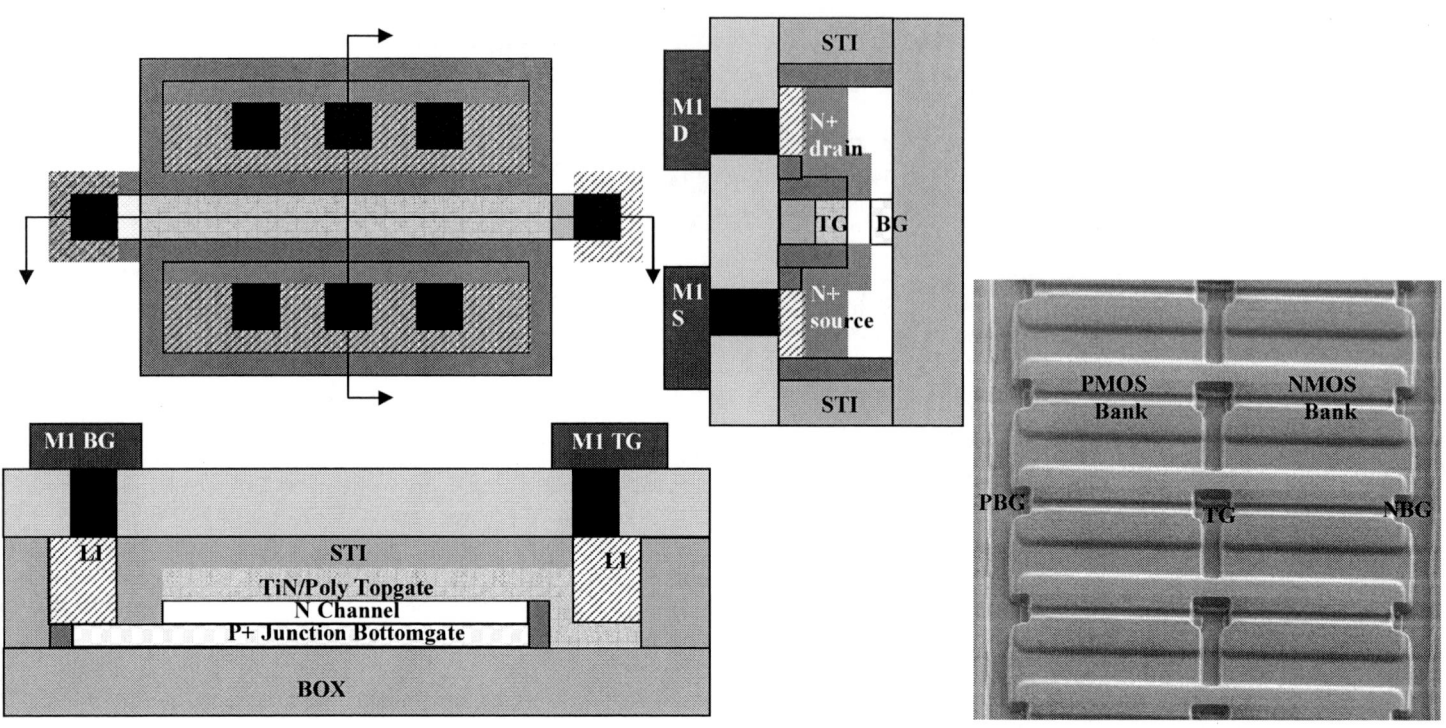

Fig. 1 FlexFET NMOS Top View and X & Y Section Drawings

Fig. 2 FlexFET IDG-CMOS Inverter Oscillator SEM

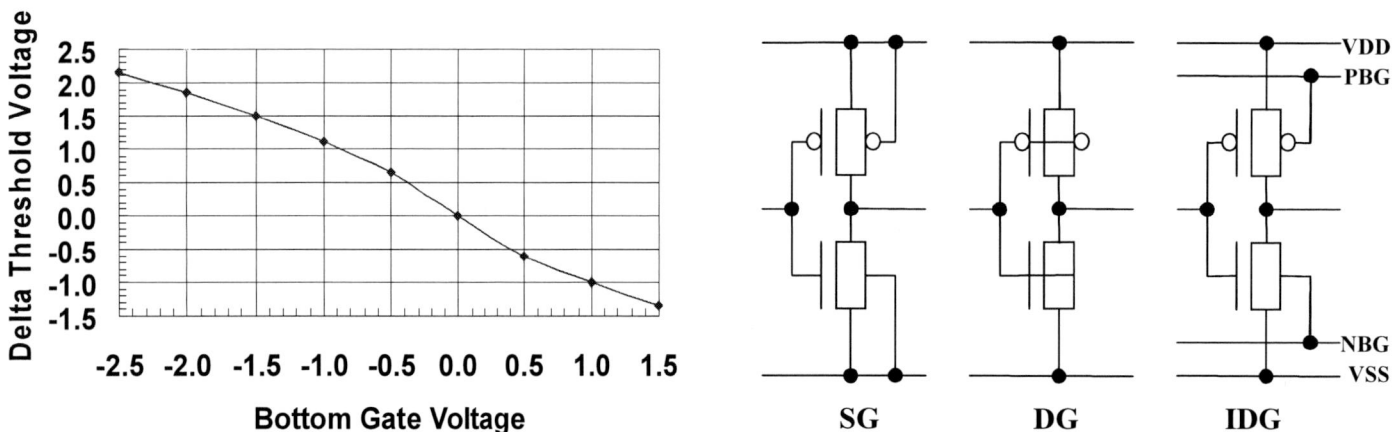

Fig. 3 Measured NMOS Vt shift vs. BG Voltage, showing >0.85V/V

Fig. 4 SG, DG, IDG Ring Oscillator Schematics

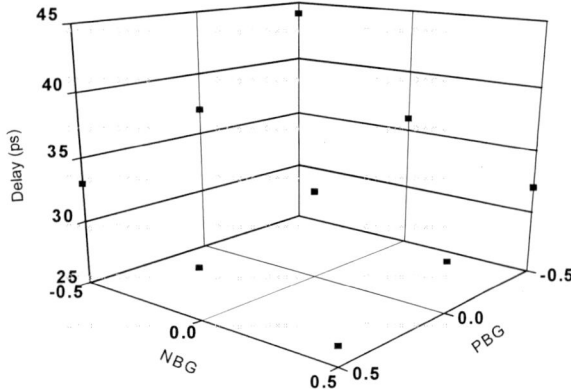

Fig. 5 IDG-CMOS Propagation Delay vs. BG Voltages

		Slow ULP	Nominal	Fast HP
NBG & PBG	V	-0.5	0	+0.5
VT	V	0.85	0.425	0
Ioff	uA/um	1.0E-10	1.0E-05	1
Pstandby	uW/um	2.5E-10	2.5E-05	2.5
Ion (NMOS)	uA/um	520	700	850
Pactive	uW/um	191	263	323
Prop Delay	psec	44	32	26
PDP	fJ/um	8.4	8.4	8.4

Fig. 6 180nm FlexFET IDG-SOI CMOS Performance Table

Secondary Ion Mass Spectrometry Analysis of Wafer Contamination Resulting from Gloved Hands

Wendy Morinville and Chantelle Krasinski
Micron Technology, Inc., Surface Analysis Laboratory
Boise, Idaho 83716

Abstract---**It is well known that wafer-handling protocols to avoid contamination are a mandatory part of semiconductor fabrication. One of the most important pieces of the standard "cleanroom" suit is the glove worn to eliminate contamination from human skin, of which mobile ions from fingerprint oils are a major contributor. This paper will show that wafer handling, even with fabrication-recommended gloves in place, can still significantly contaminate the wafer.**

Keywords-contamination, secondary ion mass spectrometry, gloves

I. Introduction

Many precautions to reduce human contamination in semiconductor manufacturing areas are utilized, including restricting the use of makeup, lotion, etc., as well as wearing appropriate cleanroom suits and gloves. It would seem reasonable that using gloves would effectively eliminate mobile ion contamination from the hands, since gloves form a protective barrier to keep oils on the skin from coming into contact with the silicon surface. However, what if the gloves themselves are a significant source of mobile ion contamination?

Typically, analysis of gloves is done by a digestion and subsequent chemical analysis. Although this technique provides contaminant information within the gloves, it does not adequately represent the extent to which those elements transfer to the surfaces with which they come into contact. It was found during the course of this experiment that the gloves themselves contain many of the same elements present in human contamination and that these elements can be transferred to the wafer surface.

II. Experimental Method

Two independent SIMS experiments were conducted using a Perkin Elmer 6300 quadrupole SIMS instrument. In study A, a 2 keV O_2 primary ion beam was rastered over a 500μm x 500μm area, and the positive ions from the central 150μm x 150μm area were collected. In study B, 1 keV O_2^+ and 1 keV Cs^+ primary ion beams were rastered over an 800μm x 800μm area, and the positive and negative ions, respectively, from the central 280μm x 280μm area were collected.

For each phase of the study, one prime silicon wafer was broken into several pieces. The sample preparation area and tools used were cleaned with isopropyl alcohol between each test to prevent any cross contamination. Two "control" areas of each wafer were reserved: one piece was not touched and one piece was touched with bare hands. The remaining samples were touched using various gloves certified for cleanroom use. In all cases, great care was taken to avoid allowing the outside of the gloves to contact any surface. The following gloves were utilized for study: latex, latex acid, three nitrile types [A, B, and C], and vinyl.

III. Results and Discussion

In study A, the wafer pieces were first touched in a single tapping motion while wearing each type of glove. Fingerprints were visible on all of the wafer pieces after handling. Analysis was taken on three regions of each fingerprinted area and the average results shown normalized to the untouched control (Fig. 1). The samples were also analyzed in reverse order, after silicon sputter, to verify that trends observed were not due to SIMS memory effects [1].

Samples were next prepared by touching the gloves to the wafer pieces while handling the glove with clean tweezers only. This was done to verify that the contamination was due to the gloves, and not to oils diffusing through the gloves from the skin. An interesting observation from the two preparation methods was that higher contamination levels were observed on the wafer after the glove was applied with tweezers. One possibility is the pressure dependence suggested by Hartzell *et al.* i.e., that larger force was applied between the glove and sample when using the tweezers [2]. These findings also confirmed that the contamination is originating from the gloves and not diffusing through the gloves.

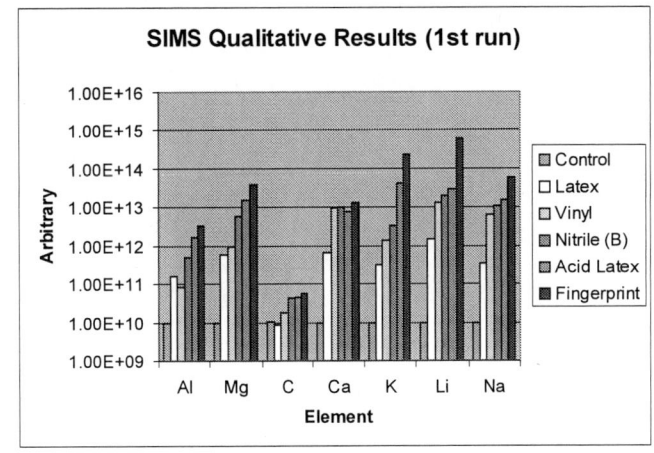

Figure 1: SIMS qualitative results from study A.

In study B, the experiment was repeated with the addition of two new types of nitrile gloves. The results trended with study A (Fig. 2) and differences between studies are within the variation range expected due to the inhomogeneous nature of the residue remaining on the wafers [3].

35

1-4244-0374-X/06/$20.00 © 2006 IEEE

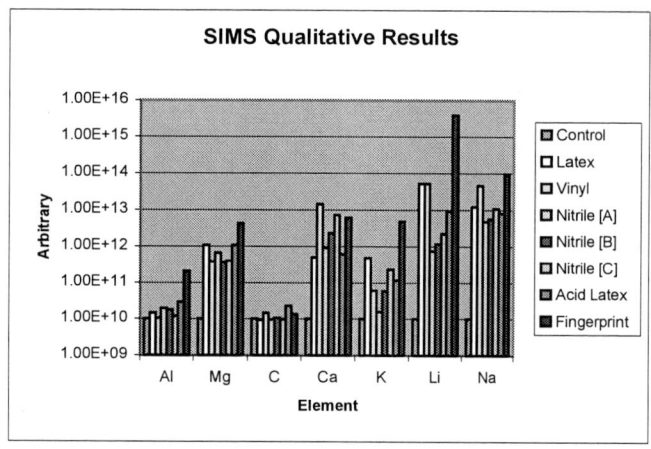

Figure 2: SIMS qualitative results from study B.

The fingerprints were also analyzed using auger electron spectroscopy (AES) to provide the particle distribution and elemental mapping. In Fig. 3, the vinyl glove print is clearly shown. The AES elemental map (Fig. 4) at a higher magnification shows carbon, calcium, and chlorine. The AES results are consistent with the SIMS findings that the vinyl glove had high levels of contamination. AES images for the nitrile [A] glove (Fig. 5) show considerably less residue correlating to the lower levels of contamination observed by the SIMS for that glove.

Figure 3: AES image of vinyl print.

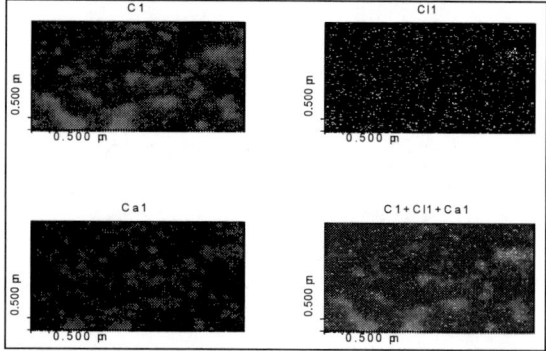

Figure 4: AES map of vinyl print corresponding to higher magnification image of Figure 3.

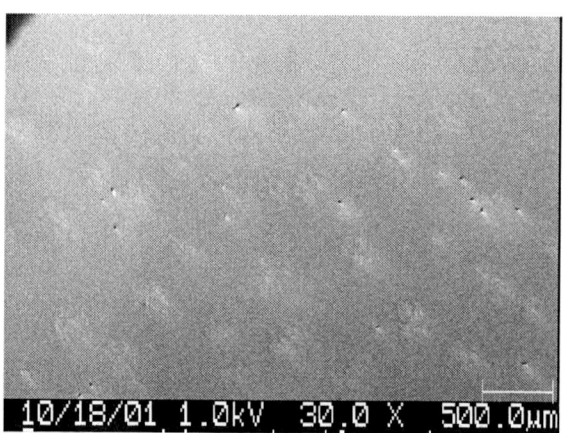

Figure 5: AES image of nitrile [A] print.

IV. Conclusion

It has been shown that significant mobile ion contamination may be added during wafer handling, even if the appropriate gloves are used. The contamination levels transferred to the wafers from gloves can be comparable to those observed from bare hands and similarly capable of skewing analysis results or affecting the process. Unlike airborne particulate contamination, merely avoiding contact between the gloves and wafer can eliminate the contamination from direct contact with the gloves. As a final remark, it is strongly suggested that silicon wafers be handled using the following protocol during manufacturing, or when preparing samples for surface analysis or device testing:

1) Wafers should only be handled with wafer handling wands, tweezers, or wafer handlers.

2) If wafer cleaving is necessary, cleave while handling the sample with wafer tweezers and a sharp diamond scribe on a clean sample preparation station.

3) When handling the samples with tweezers, the area to be analyzed or tested should be avoided.

4) Do not touch the cleaved sample or sample holder if possible. Knock off any particles that have appeared on the sample by tapping.

5) Never use tape to adhere the wafer or cleaved samples to a flatpack or other sample transportation device. The reason is two-fold: increased contamination from the tape and increased handling to remove the sample if taped down.

Acknowledgements

The authors would like to thank Harold Krasinski, Micron Surface Analysis Lab, for the AES analysis.

References

[1] R.G. Wilson, F.A. Stevie, and C.W. Magee, *Secondary Ion Mass Spectrometry*, (Wiley, New York, 1989) p. 2.8-1.

[2] A. Hartzell, J. Rose, D. Liu, P. McPherson, M. O'Shaughnessy, C. Seeley, R. Burt, Micro **October**, 69, (1996).

[3] S. P. Smith, J. Metz, and P. K. Chu, *Secondary Ion Mass Spectrometry; SIMS XI Proceedings*, (Wiley, New York, 1998), p. 233.

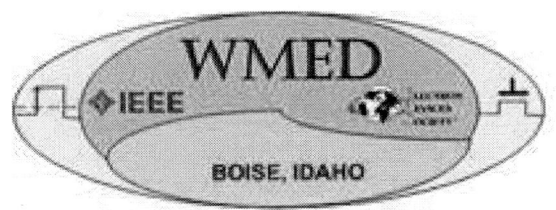

Session

Poster Papers and Abstracts

THERMAL NOISE LIMITS IN NANOSCALE ELECTRONICS

M. Mudrow, W. Wanalertlak and L. Forbes
School of Electrical Engineering and Computer Science
Oregon State University, Corvallis, OR 97331-5501
Email: prof@forbes4.com Phone: (541) 753-1409

Abstract-- **An analysis of thermal noise predicts the number of bit errors per year caused by noise in nanoscale electronic memories and processors will be excessive. Noise it seems has already imposed a fundamental limit on the speed of microprocessors. No significant advance in microprocessor speeds is anticipated due not only to power limitations but also noise limits.**

*Index terms—***bit error rates, logic error rates, memory error rates, thermal noise**

I.   INTRODUCTION

An analysis has previously been made of the increasing portion of the threshold voltage being occupied by thermal noise levels and the bit error rates in digital logic circuits [1]. The use of charge storage on nanoscale particles has been realized in memories [2,3]. Nanoscale memories allow memory density to be scaled down significantly. It is intuitively obvious that at very small memory sizes the capacitive charge in the element is very small and thermal noise levels will have a larger impact on the leakage current in the storage element. It is important to recognize this fundamental problem and understand the implications for bit error rates in memory circuits.

II.   NANOSCALE LOGIC

Rice's formula gives the mean frequency of estimated crossings of a threshold in a stationary Gaussian process with zero effective value [4]. This method has been used to analyze the effect of thermal noise in integrated circuits with sizes below 40nm.

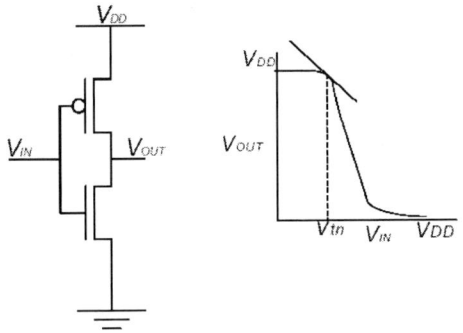

Fig. 1. CMOS inverter showing threshold for bit errors.

For the CMOS inverter in Fig.1 the bit error rate, BER, is

$$BER = \frac{2}{\sqrt{3}} f_c t N e^{-V_{th}^2/(2V_n^2)} \qquad (1)$$

This has been used [1] to calculate the bit error rate in nanoscale CMOS logic circuits due to thermal noise.

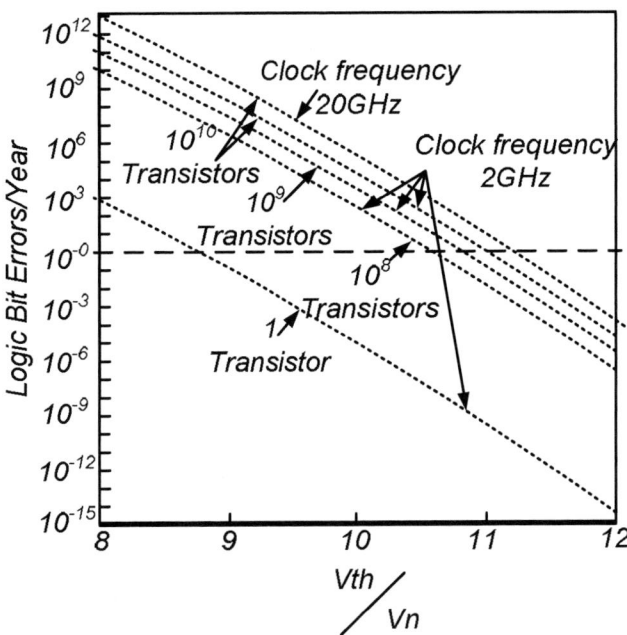

Fig. 2. From Kish[1], bit error rate in CMOS logic due to thermal noise.

III.   NANOSCALE MEMORY

Figure 3 indicates a 3 nm particle embedded in silicon oxide over a CMOS transistor channel [2]. This particle has a capacitance associated with that can be evaluated from Poisson's Equation evaluated for a sphere over a plane resulting in

$$C = 4\pi\varepsilon r = 4.45\times10^{-19} \qquad \text{F} \qquad (2)$$

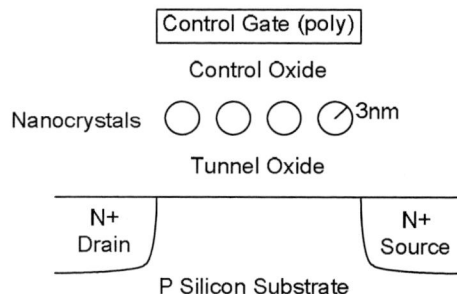

Fig. 3. Nanoscale nonvolatile memory.

Assuming a reasonable retention time of about 1 year for the memory, then the resistance associated with the leakage current [5] in the transistor gate tunnel oxide is given by:

$$\tau = RC = 3.2 \times 10^7 s = R = 7.2 \times 10^{25} \Omega \qquad (3)$$

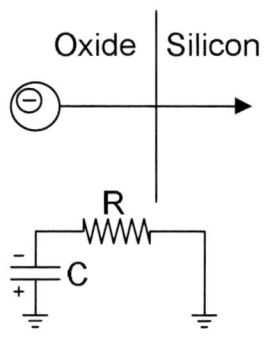

Fig. 4. Tunneling resistance and thermal noise.

This R-C combination in Fig. 4 results in a corner frequency of $f_c = 1/(2\pi RC) = 5 \times 10^{-9}$ Hz. The noise associated with tunneling currents is primarily white noise and the $1/f$ noise can be ignored [6]. The thermal noise of the tunneling resistor can then be calculated using the "brick wall approximation [7]."

$$\int \overline{i_n^2} df = \frac{\pi}{2} f_c 4kT \frac{1}{R} \qquad \mathrm{A}^2 \qquad (4)$$

$$\sqrt{\overline{i_n^2}} = \sqrt{\frac{kT}{C}} \times \frac{1}{R} \qquad \mathrm{A_{rms}} \qquad (5)$$

$$\sqrt{\overline{v_n^2}} = \sqrt{\frac{kT}{C}} = \sqrt{\frac{(1.6x10^{-19}C) \times (0.025V)}{4.45 \times 10^{-19} F}} = 0.096 \quad \mathrm{V_{rms}} \qquad (6)$$

Rice's formula can also be applied to a memory element to determine the bit error rate caused by the thermal noise.

$$BER = \frac{2}{\sqrt{3}} f_c t N e^{-V_{th}^2/(2V_n^2)} \qquad (7)$$

BER is the bit error rate, f_c is the corner frequency, t is the time, N, is the number of memory elements, V_{th} is the noise threshold voltage, and V_n is the effective noise voltage seen on the resistor. To calculate a theoretical bit error rate in Fig. 5 we assume a memory size of 1G bit ($N=10^9$), $t=3.2x10^7$ s, $fc=0.5x10^{-9}$ Hz, $V_n=.096$ V, and the noise threshold voltage given by half the charge on the capacitor assuming a 1 V supply ($V_{th}=0.5$ V). Eq. 7 results in a bit error rate of approximately 10^2 bit errors per year at 85°C.

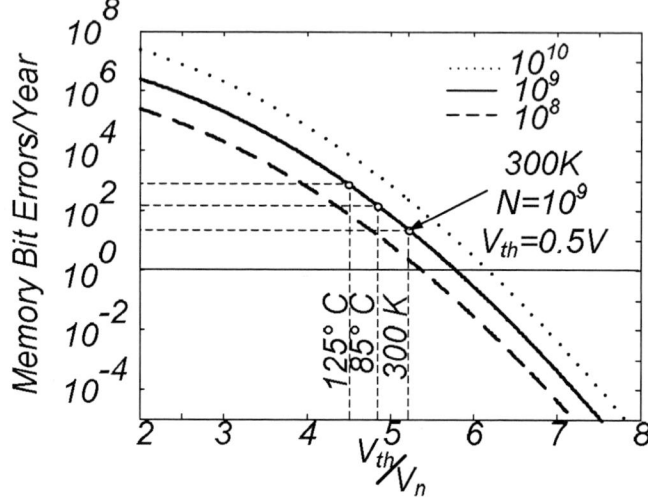

Fig. 5. Calculated bit error rate in nanoscale memory.

IV.   CONCLUSIONS

Noise has it seems already imposed a fundamental limit on the speed in microprocessors.[1] No significant advance in microprocessor speeds is anticipated. The bit error rate in nanoscale memories is high compared to the one bit error per year of current memory technologies [8].

This work explores and describes the fundamental limits imposed on nanoscale electronic devices by noise.

References
[1] L.B. Kish, IEE Proc.-Circuits Devices Syst., vol. 151, no. 2, pp. 190-194, April 2004.
[2] Muralidhar et al., IEDM Digest, 2003, pp. 601-604.
[3] B. Hradsky et al., IEEE Device Research Conf., 2005, pp. 37-8.
[4] S.O. Rice, Bell Syst. Tech. J., vol. 23, pp. 282-332, 1944; and vol. 24, pp. 46-156, 1945.
[5] M. She, Y.-C. King, T.-J. King and C. Hu, IEEE Device Research Conf., 2001, pp.139-140.
[6] T.G.M. Kleinpenning, Solid-State Electronics, vol. 21, pp. 927-931, 1978.
[7] D.A. Johns and K. Martin, *Analog integrated Circuit Design*, J.Wiley & Sons, N.Y., 1997, pp. 180-220.
[8] J. Ziegler, M. Nelson, J. Shell, R. Perterson, C. Gelderloos, H. Muhlfeld, and C. Montrose, IEEE J. Solid-State Circuits, vol. 33, pp. 246-252, Feb. 1998.

A Novel Triple Mode LNA Designed in CMOS 0.18um technology for Multi Standard Receivers

M.B. Vahidfar[1], O. Shoaei[1]

[1]IC Design Center, ECE Department, University of Tehran

m.vahidfar@ece.ut.ac.ir

Abstract— **A novel reconfigurable, area optimized low noise amplifier designed for DCS1800, UMTS and IEEE802.11b/g standards is described in this paper. The design is based on enhanced single stage common gate architecture. Two feedback loops are employed to LNA be easily reconfigured for each standard. These feedback loops not only enhance the input matching and NF parameter but also improve the linearity of the LNA. A switched capacitor resonator is responsible for the band selection. The design is done in a 0.18um CMOS technology and the simulation results show that the NF and S11 parameters are better than 2.3dB and -17 dB respectively in each frequency band. The design meets the performance requirements of above mentioned standards. The design also achieves low area and low power consumption in comparison with a single band LNA.**

***Key words*: Low noise amplifier (LNA), CMOS RF, multi band, common gate, WLAN, UMTS, DCS, Multi standard.**

I. INTRODUCTION

Nowadays, the demands for multi-standard receivers which support various wireless standards are rapidly increasing. The cost effective, fully integrated multi standard receivers can only be achieved by developing reconfigurable RF front ends which means blocks that can be used for any standard of any frequency band. A novel programmable CMOS LNA which is area optimized and works in triple bands is presented in this paper.

II. MULTI STANDARD LNA

A. CG Architecture for multistandard applications

The main advantage of Common gate (CG) architecture is its resistive input impedance for impedance matching, but due to its high NF, it is not applicable in many cases [3].

B. CG LNA enhansment using voltage-voltage feedback

Scaling the CMOS technology to sub micron provides transistors with higher FT which makes possible the use of feedback loops in Giga-hertz frequencies to achieve reconfigurability besides lower noise and higher linearity. As it is shown in figure 1 and it is discussed in this section, an enhanced CG LNA can be achieved by employing a voltage-voltage feedback loop around the LNA [1]. Considering figure1 and feedback theory, it can be

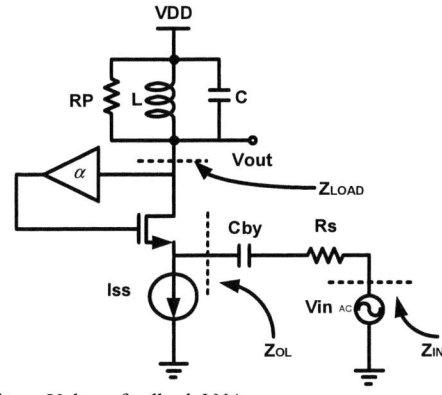

Fig. 1. Voltage-Voltage feedback LNA

shown that the input impedance of LNA in closed loop condition is:
$$Z_{in} = R_s + \frac{1}{g_m} + \alpha.R_P \qquad (1)$$

Assuming the input impedance matching condition:
$$R_s = \frac{1}{g_m} + \alpha.R_P \qquad (2)$$

In the other words equation (2) shows that if the output resonator is multi band, the narrow band input matching can be achieved by a feedback loop without using any other resonator in the input. The feedback loop also can be implemented easily by a capacitor divider as it is shown in figure 2.

C. Noise Figure

The thermal noise of the MOS transistor has the main contribution in the total NF of this LNA which can be modeled by:
$$\overline{i_{n\_channel}}^2 = 4KT\gamma g_{d0} \qquad (3)$$

Where g_{d0} is the zero-bias drain conductance of the device, and γ is a bias dependent factor [2]. The contribution of this noise source on total LNA NF can be formulated by:

$$NF1 = \frac{4}{g_m R_s} \frac{\gamma}{\alpha} \frac{1}{\left[\left(1 + \frac{g_{mb}}{g_m} + \frac{1}{g_m R_s} \right)^2 + \left(\frac{C_{gs}}{C_2 + C_{gs}} \right)^2 \left(1 + \left(\frac{\omega_0 C_2}{g_m} \right)^2 \right) \right]}$$

In which ω_0 is the working frequency, and α is defined as follows: $\alpha = \dfrac{g_m}{g_{d0}}$

In spite of the CG architecture, in the enhanced CG LNA, the NF is optimized by the bias current and input matching is achieved by adjusting the feedback factor. Moreover this topology achieves better performance in NF comparing to CG LNA, due to $g_m R_s \gg 1$.

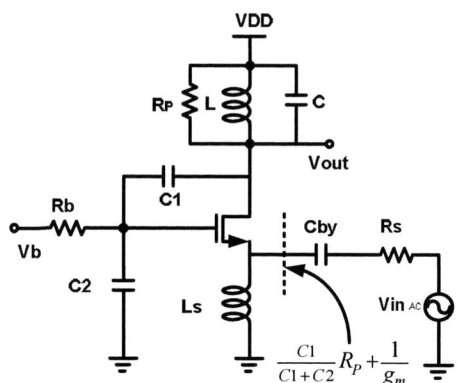

Fig. 2. Implementation of voltage-voltage feedback LNA

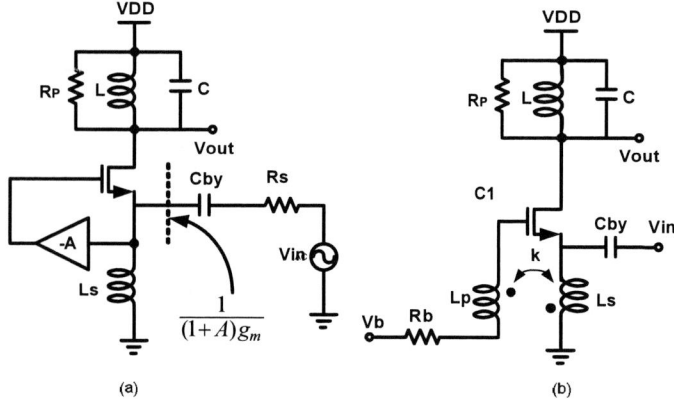

Fig. 3. a) Gm-boosting CG architecture. b) Implementation

D. Gm-bossted LNA

As it was discussed in section C, the NF enhancement is achieved by increasing the LNA bias current which leads to higher gm. However the required NF can be achieved by employing another feedback loop in odder to increase gm without increasing the LNA bias current. Referring to figure 3, the transconductance of this LNA is enhanced by a factor of (1+ A) comparing to conventional CG LNA. It can be shown that the NF of this architecture caused by thermal noise is also decreased by square of (A+1) factor [5]. The main drawback of this technique is degrading S11 parameter by reducing input LNA resistance. The gm-boosted LNA can be implemented by cross-coupled technique in a differential LNA [5] and by on-chip transformer in a single ended LNA [4].

III. TRIPLE BAND LNA

As it is shown in figure 4, the designed triple band LNA is made by employing both of voltage-voltage feedback and gm-boosted techniques in order to meet the input matching and NF requirements independently. When the NF is optimized by adjusting the gm, the required input impedance matching is achieved by C2 capacitor adjustment. The programmable output resonator is made by using an inductor and two switched capacitors. The load switching is achieved by SWb and SWc switches which implemented by NMOS transistors and are biased by a small PMOS transistor when they are off in order to reduce their parasitic capacitors. The switches are optimally sized to alleviate the trade off between the parasitic capacitors of the switches and their contribution in the LNA output noise. The biasing condition of LNA is programmed for each mode to achieve NFmin. The LNA works in each mode of DCS, UMTS and IEEE802.11b/g by reconfiguring the output resonator and biasing condition.

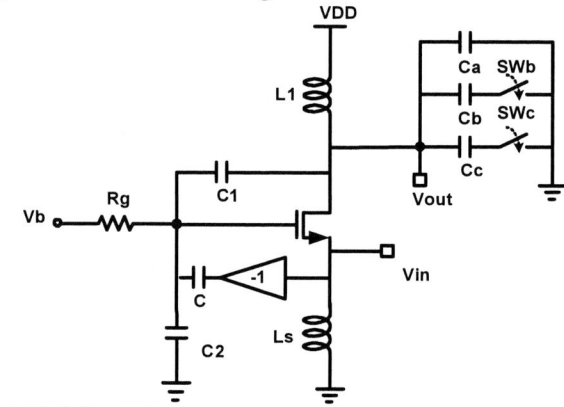

Fig. 4. Triple band LNA

IV. SIMULATION RESULT

The simulation was done in a 0.18 um CMOS technology. The Q of the load inductors and SMD inductor which is off-chip are about 7 and 15 respectively. The bad effect of bond wire inductor on VDD pin is reduced by placing 3 pins in parallel. The worst cases of NF, S11, S21 and IIP3 in three frequency modes are 2.3dB, -17dB, 10dB and 19dBm respectively. The LNA consumes 10mA from a 1.5V supply.

REFERENCES

[1] Paolo Rossi, Antonio Liscidini, Massimo Brandolini and Francesco Svelto, "A variable gain RF front end based on a voltage-voltage feedback LNA for multi-standard applications" IEEE J. Solid-State Circuits, Vol. 40, NO. 3, March 2005.

[2] D. K. Shaeffer, T. H. Lee, "A 1.5 V, 1.5 GHz, low noise amplifier" IEEE J. Solid-State Circuits, vol. 32, NO. 5, May 1997.

[3] H. Hashemi,A. Hajimiri, "Concurrent Multiband Low-Noise Amplifiers—Theory, Design and Applications", IEEE Transaction on microwave theory and techniques, Vol. 50, NO. 1, January 2002.

[4] Xiaoyoung Li, et all, "Low Power gm-boosted LNA and VCO circuits in 0.18um CMOS", ISSCC, Session 29, 2005.

[5] D. Allstot et al., "Design Considerations for CMOS Low-Noise Amplifiers", RFIC symposium, pp. 97-100, 2004.

POWER DISSIPATION AND TEMPERATURE VARIATIONS IN NANOSCALE DEVICES

W. Wanalertlak, M.Y. Louie, and L. Forbes
School of Electrical Engineering and Computer Science
Oregon State University, Corvallis, OR 97331-5501
email: prof@forbes4.com phone: (541) 753-1409

Abstract—**Equivalent circuit techniques are used to obtain time and frequency dependent solutions to the diffusion equation. These are demonstrated to provide detailed time and frequency dependent temperature variations in nanoscale electron devices.**

Index Terms—**diffusion equation, nanoscale devices, power consumption, temperature variations**

I. INTRODUCTION

The transition to nanoscale devices implies much higher power densities. Fig. 1 illustrates the past and projected history of CMOS devices in integrated circuits. The 1990 dimensions were about 1 micrometer or 1000 nanometers with transistor currents of around 1mA/um and 5V power supplies. For a minimum dimension device, 1um x 1um, this translates into $5x10^5$ W/cm^2. Typical device currents have remained around 1ma/um while devices have scaled down, in the 2000 time frame 0.1um, 100nm, devices used 1.5 V power supplies and the projection into the 2010 time frame is 0.01um or 10nm dimensions with 0.5V power supplies which results for a minimum dimension device in power densities of $5x10^6$ W/cm^2. Such power densities will result in very high device temperatures even with ideal heat sinks and advanced cooling techniques, many not even yet developed.

II. EQUIVALENT CIRCUIT TECHNIQUES

What is new here is our treatment of time and frequency dependent solutions to the diffusion equation by transmission line techniques and the application to nanoscale devices[2,3]. The diffusion equation in rectangular coordinates describing heat conduction [1] is;

$$a \frac{\partial^2 T}{\partial x^2} = \frac{\partial T}{\partial t} \qquad (1)$$

where a is the thermal diffusivity, m^2/sec in MKS units and T is the temperature.

This is the same type of equation as that describing voltages on a distributed R-C transmission line in electric circuits

$$\frac{\partial^2 V}{\partial x^2} = R C \frac{\partial V}{\partial t} \qquad (2)$$

An equivalent circuit representation can be made in rectangular coordinates for heat conduction as shown in Fig. 2 where for each volume element,

$$R = 1/K A \qquad\qquad K/W\,m \qquad (3)$$
$$C = C_p\, \rho\, A \qquad\qquad J/K\,m \qquad (4)$$

where K is the thermal conductivity, C_p the heat capacity, ρ the density, and A the area of the sample through which there is heat conduction. Temperature is analogous to voltage and heat flux analogous to current. These R-C transmission lines have been previously analyzed [2, 3] and are diffusion lines where potential and currents are described by the diffusion equation.

We can obtain time and frequency dependent solutions to the diffusion equation by a transmission line analysis of Fig. 2, [2,3] and apply these to the temperature of nanoscale electron devices. Here we then investigate temperature variations in nanoscale devices, and demonstrate both transient time dependent and frequency dependent variations in the temperature of nanoscale devices.

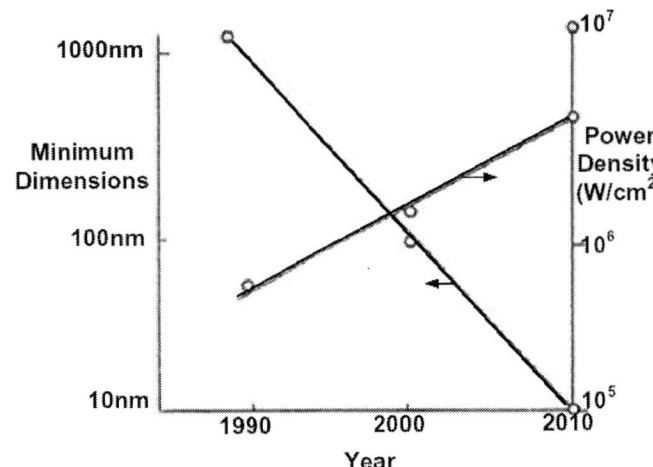

Fig. 1. Increase in power density in integrated circuits with time and decreasing feature sizes.

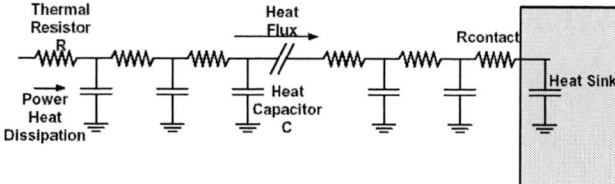

Fig. 2. Equivalent circuit for heat flow.

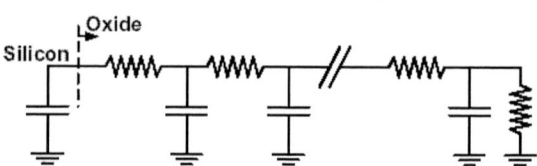

Fig. 4. Equivalent circuit for heat flow and temperature of the silicon on insulator transistor, the R-C transmission line represents semi-cylindrical elements of the oxide or insulating substrate.

III. PRACTICAL EXAMPLE

As an illustration of a practical problem we consider a silicon on insulator transistor as shown in Fig. 3. This transistor has a sinusoidal signal applied and an average power dissipation of 50 μwatts. This might well be representative of a low power silicon RF power amplifier and transmitter. The device has a channel length of 100nm and width of 100nm. The power is assumed to be dissipated in half of a cylinder with radius 50 nm and the heat flow is to a heat sink at the back of the substrate wafer 0.1cm away. Fig. 4 shows the equivalent circuit and Fig. 5 the results. The bottom part of Fig. 5 shows the $\sin^2(\omega t)$ power dissipation at 100 MHz in this case with an average value of 50 μwatts. The high frequency variations in temperature of the transistor are attenuated by the heat capacity of the oxide, however, there is a steadily increasing value of the average temperature over a 200ns time period.

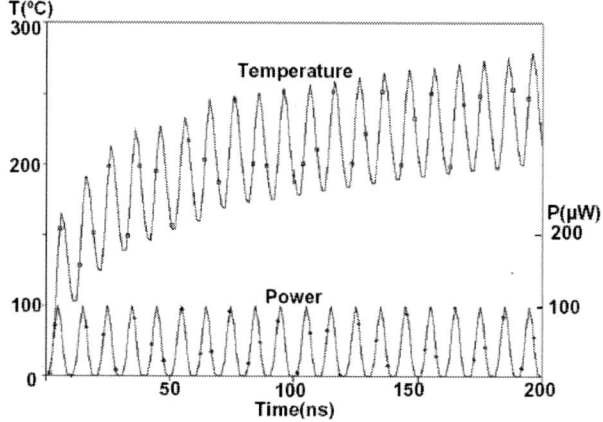

Fig. 5. Illustration of the time dependent increase in temperature of the low power RF silicon on insulator transistor .

IV CONCLUSIONS

A textbook formulation of the problem of frequency dependent solutions to the diffusion equation has been given by Kittel and Kroemer [1]. They give no solution to this extremely difficult differential equation and note only that high frequency variations will be strongly attenuated. The equivalent circuit techniques allow solutions to be readily obtained by circuit simulations.

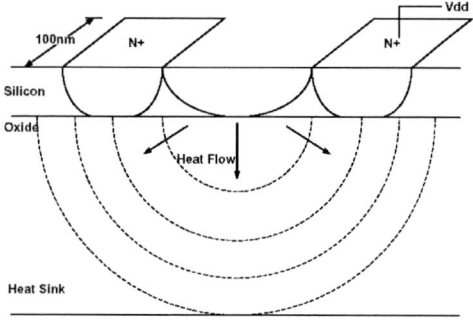

Fig. 3. Silicon on insulator low power transistor with a 50μwatt power dissipation at 100 MHz.

REFERENCES

[1] C. Kittel and H. Kroemer, Thermal Physics, 2nd Ed., New York: Freeman, p. 424, 1980.

[2] L. Forbes and M.S. Choi, "l/f noise due to temperature fluctuations in heat conduction," Ext. Abst. Device Research Conference, Santa Barbara CA, 1996, pp. 46-47.

[3] L. Forbes, M.S. Choi and W. Cao, "1/f noise in bipolar transistors due to temperature fluctuations in heat conduction," Microelectronics Reliability, vol. 39, no. 9, pp. 1357-1364, Sept. 1999.

Micro-Sensor for Monitoring Oils

Brian M. Marx, Matthew Luke, and Darryl P. Butt, *Department of Material Science and Engineering, Boise State University, Boise, ID*

Abstract—An electrochemical sensor has been constructed for detecting changes in the electrical responses of oil/metal systems. The sensor was used to make Electrochemical Impedance Spectroscopy (EIS) measurements and the impedance responses are compared as a function of oil additive. This work uses the base stock oil (no additives) as the baseline for comparison of the various additives. Of the four additives tested, only the metal deactivator additive was found to increase the impedance. The full formulation oil (containing all additives) displays the lowest impedance, indicating possible additive interaction. A basic Randle's circuit was used to fit the impedance data. This type of fit provides information pertaining to the properties of the bulk oil and contributions due to the interactions at the oil/metal interface.

Index Terms—electrochemical sensor, electrochemical impedance spectroscopy, oil, additive

I. INTRODUCTION

FOR any mechanical system containing rotating hardware, where moving parts are in contact, lubricants must be used to maintain the integrity of the system. Despite the increasing use of solid lubricants, oil and grease remain the most popular lubricants in mechanical systems. Many studies have focused on measuring the performance of oils through empirical experiments. However, very little information exists concerning the properties of the oil, particularly at the oil/metal interface. For this reason, a micro-sensor is being developed for analyzing oils through the use of Electrochemical Impedance Spectroscopy (EIS). This micro-sensor will be capable of detecting oil degradation and/or contamination (i.e. detecting the presence of water or other fluid in the system). In addition EIS is a powerful technique for analyzing metal/liquid interfaces and the data can be used to describe the oil additive effects at these interfaces. Additive effects will be the focus of this paper.

A variety of functional additives are used in Aerospace Ester-based lubricants to provide improved oxidative thermal stability, wear and corrosion inhibition, as well as other important lubricant properties through prescribed additive formulations. Additives include anti-oxidants, anti-wear additives, corrosion and rust inhibitors, and anti-foaming agents. In addition, organics such as alcohols and fatty acids may exist in the lubricant system or be produced in service. Minor additive components of the lubricant are adsorbed on metal surfaces and have both beneficial as well as detrimental effects on the interfacial behavior of metal parts. In this phase of the research, a systematic study is carried out to assess the effects of additives and environmental by-products.

II. DEVICE FABRICATION

While there is a wealth of information pertaining to the use of EIS for studying aqueous systems and coatings, very little information exists concerning the application of EIS on oil systems [1-8]. A simple EIS sensor design used by some workers is comprised of two-interlocking "combs" and is used to increase the capacitive surface area in a micro-device[8]. Because this design increases the measurement accuracy (due to high surface area) while minimizing the overall geometric size, this design was chosen as a good candidate for an oil sensor. The two "combs" function as electrodes and the spacing between them is small due to the high resistance of the oils being tested (GΩ range). A 250 μm spacing was used to separate the electrodes due to the limitations of the screen-printing process used here. A picture of the sensor, with leads connected, is shown in Fig. 1. The substrate is a LTCC ceramic and the electrodes are silver. The dog-bone shape is for a different application and will not be discussed in this article.

Figure 1 The constructed sensor with leads connected for making impedance measurements. A blow-up of the "comb" design shows the expanded capacitive area of the sensor.

Manuscript received February 17, 2006. This work was supported by Pratt and Whitney.

B.M. Marx is with the Dept. of Material Science at Boise State University (208-426-5376; fax: 208-426-2470); e-mail: brianmarx@boisestate.edu).

M. Luke is an undergrad with the Dept. of Material Science at Boise State University (e-mail: mattluke@boisestate.edu).

D.P. Butt is with the Dept. of Material Science at Boise State University (e-mail: darrylbutt@boisestate.edu).

III. Experiment

Impedance measurements were made with a Gamry EIS 300 system and an EG&G/Perkin Elmer impedance system using PowerSuite software. All experiments were performed using a two electrode set-up in a Faraday cage at room temperature. The sensor was placed in the various oils, with and without additives, and the open circuit potential (OCP) was measured. Once the OCP stabilized, impedance measurements were made at the OCP. A 10 mV AC perturbation voltage was applied to the OCP and a frequency scan was performed from 100000 Hz to 0.01 Hz.

IV. Results and Discussion

The sensor performed very well in detecting differences in the impedance for the various additives as shown in the Nyquist plot in Fig. 2. Single additives were mixed with basestock oil to create the oil formulations. The oil number (used to track the oil mixtures) and additive type are shown in Fig. 2. It can be seen that the BP full formulation oil (containing all additives) has the lowest impedance of the oil formulations. This would suggest that there is some interaction between the additives and the interaction needs to be explored further (i.e. mix additives in set ratios and measure the changes in the impedance response). The extreme pressure additive decreases the impedance relative to the base stock, while the metal deactivator increases the impedance beyond that of the base stock. This would indicate that the metal deactivator additive coats the metal surface and this "coating" is more resistive than the base stock. Some surface analysis work will be carried out to determine the chemistry of the metal/oil interface.

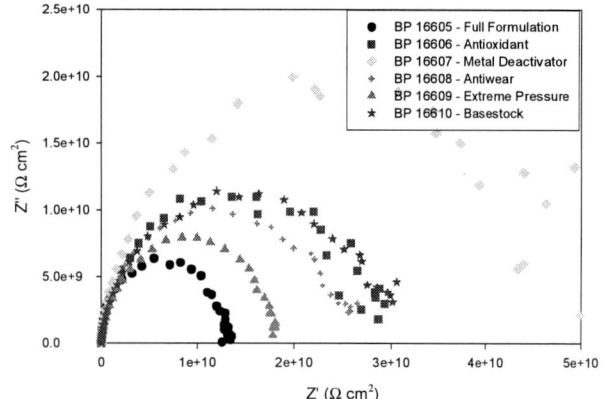

Figure 2 Nyquist plot for various oil additives. The numbers provided in the plot legend represent the additive/basestock oil formulation.

Basic impedance analysis was carried out using a simple Randle's circuit (a parallel RC circuit in series with a resistor) and the results are shown in Table 1. R_{soln} represents the resistance as related to the bulk oil, R_{CT} is the resistance due to charge transfer at the interface, and C_{dl} accounts for the double-layer capacitance [4]. These results show that the resistance to charge transfer at the oil/metal interface is a few

orders of magnitude larger than that of the bulk oil. Surface studies will be performed to determine what reactions are occurring at the interface. By understanding the chemistry of the interface, a more representative circuit model can be constructed and used to interpret the data.

Table 1 The fit parameters yielding the best fit to the impedance data as modeled by a simple Randle's circuit. The number corresponds to the additive as shown in Fig. 2.

	R_{CT} (Ω cm^2)		R_{soln} (Ω cm^2)		C_{dl} (F)	
	value	error	value	error	value	error
16605	1.3E+10	7.1E+07	5.3E+04	2.3E+04	1.1E-11	4.6E-14
16606	2.5E+10	1.5E+08	5.9E+04	2.0E+04	1.4E-11	5.4E-14
16607	4.0E+10	2.6E+08	6.4E+04	1.9E+04	1.5E-11	5.7E-14
16608	2.2E+10	1.3E+08	5.7E+04	2.2E+04	1.3E-11	5.1E-14
16609	1.7E+10	9.7E+07	5.7E+04	1.9E+04	1.5E-11	5.9E-14
16610	2.4E+10	1.5E+08	6.2E+04	2.0E+04	1.4E-11	5.5E-14

V. Conclusions

An electrochemical sensor with a 250 μm electrode spacing was used to make EIS measurements in oils. Experiments show that differences due to oil additives can be resolved. The results indicate that the sensor could be a powerful tool for studying and monitoring oil systems. It provides a non-destructive method for probing the condition and interaction of the oil(s) and their additives. However, further surface studies need to be performed before correlating the impedance response to physical mechanisms.

References

[1] K. N. Allahar, D. P. Butt, M. E. Orazem, H. A. Chin, G. Danko, W. Ogden, and R. E. Yungk, "Impedance of steels in new and degraded ester based lubricating oil," *Electrochimica Acta*, vol. In Press, Corrected Proof.

[2] H. Bouazaze, F. Huet, and R. P. Nogueira, "A new approach for monitoring corrosion and flow characteristics in oil/brine mixtures," *Electrochimica Acta*, vol. 50, pp. 2081-2090, 2005.

[3] L. Fraigi, S. N. Gwirc, and D. Lupi, "A thick film sensor for atmospheric corrosion testing," *Sensors and Actuators B: Chemical*, vol. 18-19, pp. 558-561, 1994.

[4] V. F. Lvovich and M. F. Smiechowski, "Impedance characterization of industrial lubricants," *Electrochimica Acta*, vol. In Press, Corrected Proof.

[5] M. F. Smiechowski and V. F. Lvovich, "Electrochemical monitoring of water-surfactant interactions in industrial lubricants," *Journal of Electroanalytical Chemistry*, vol. 534, pp. 171, 2002.

[6] M. F. Smiechowski and V. F. Lvovich, "Iridium oxide sensors for acidity and basicity detection in industrial lubricants," *Sensors and Actuators B: Chemical*, vol. 96, pp. 261, 2003.

[7] M. F. Smiechowski and V. F. Lvovich, "Characterization of non-aqueous dispersions of carbon black nanoparticles by electrochemical impedance spectroscopy," *Journal of Electroanalytical Chemistry*, vol. 577, pp. 67, 2005.

[8] S. S. Wang and H. S. Lee, "The application of a.c. impedance technique for detecting glycol contamination in engine oil," *Sensors and Actuators B: Chemical*, vol. 40, pp. 193-197, 1997.

Integrated Silicon Nanowire Diodes

Justin B. Jackson[1]; Sun-Gon Jun[2]; Divesh Kapoor[1]; Mark S. Miller[1,2]; [1]Electrical and Computer Engineering; [2]Materials Science and Engineering; University of Utah

Abstract—**IV-measurements were performed on integrated silicon nanowires grown via atmospheric vapor phase epitaxy. The silicon nanowires were grown on a n-type substrate using gold catalysts, insulated using a spin-on glass, and contacted using aluminum. Rectifying behavior and low recombination is observed in the nanowire with a forward bias.**

1. Introduction

As the scaling of current fabrication technologies begins to reach its limits, other technologies will need to be developed. One potential solution is the use of silicon nanowires for bottom-up device fabrication. Silicon nanowires have been studied in recent years for use in the manufacturing of future semiconductor devices[1]. Integrating nanowire devices into future CMOS technologies will likely require several developments, such as controlling the nanowire growth location and the subsequent integration into complex systems. Purposefully placing nanowires will allow their incorporation into a much larger system. Silicon nanowires have been grown utilizing gold catalysts in atmospheric vapor phase epitaxy systems[2]. Recent work has shown that nanowire clusters can be lithographically defined, grown, and measured in situ[3].

In this study, single silicon nanowires are lithographically placed, grown, and characterized in situ. Silicon nanowire diodes fabricated using conventional lithographic processes resulted in the structure depicted in Fig. 1.

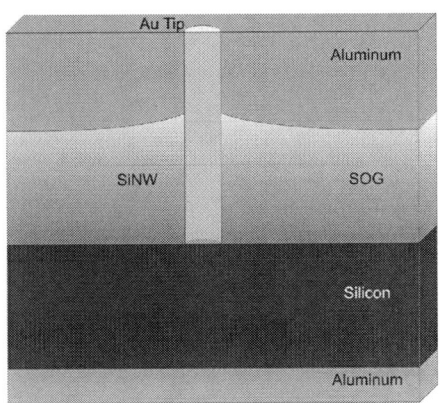

Figure 1: Representation of nanowire diode structure.

2. Growth and Patterning

N-type silicon (100) substrates were thermally oxidized to grow a 70 nm dry oxide. Arrays of 500 nm, 1 μm, 1.5 μm, and 2 μm holes were patterned on the substrate using photolithography and consequently etched using a HF buffered oxide etch. The 500 nm drawn dot sizes resulted in ~750 nm holes in the oxide. A gold thin film 10 nm thick was deposited on the surface using electron beam evaporation. The excess gold was then removed by lift off via acetone, methanol, rinsing with DI water, and blown dry using nitrogen.

Silicon nanowires were then grown by atmospheric vapor phase epitaxy by employing the vapor-solid-mechanism (VLS)[2]. Silicon nanowires were grown at 900° C for 5 minutes by $SiCl_4$ in a H_2 ambient resulting in patterned nanowire arrays. Each patterned site resulted in a dominant wire on the order of 100-300 nm in diameter accompanied by several smaller diameter wires.

The substrate was then thermally oxidized at 1000° C for 10 minutes with a 3 sccm flow of oxygen followed by a buffered HF oxide etch to remove the smaller diameter wires and also to reduce the size of the dominant wire.

3. Fabrication of Devices

To isolate the tip of nanowire from the substrate, spin-on glass was used. Accuglass T-11 311 was spun at 2000 rpm to give approximately 350 nm of oxide and subsequently heated from 250° to 425° C over 1 hour. This process was repeated to give a total oxide of ~700 nm as seen in Figure 2.

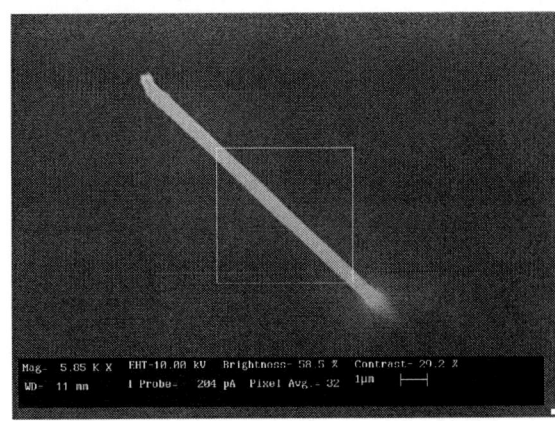

Figure 2: Silicon nanowire protruding through ~700 nm of spin-on glass.

The sample was then etched using reactive ion etching (RIE) using CF_4 and O_2 removing 85 nm of SiO_2 to expose the top of the nanowire. Aluminum, 1 um thick, was promptly deposited on to the top of the sample using electron beam evaporation. The aluminum was then patterned using an aluminum etchant at 50° C followed by annealing of the sample for 20 minutes at 450° C in N_2

ambient. Aluminum was then deposited on the back of the wafer using e-beam evaporation to give two 75 nm thick contacts on the back of the wafer followed by a 20 minute anneal at 450° C in N_2 ambient for an ohmic contact.

4. Results and Measurements

To test the quality of the manufactured diode, IV measurements were performed. Current was measured as the voltage was stepped in 20 mV increments from -1V to 1V and the resulting curve showed rectification around the origin. Also plotted were calculated 1kT and 2kT currents to show the ideality of the resulting diode. To verify that rectifying behavior occurred within the nanowire and was not attributed to a Schottky contact barrier, IV measurements were performed between contacts. These measurements also supplied information on the contact resistance of the pads.

Initial measurements were taken of the resulting nanowire structure prior to depositing of the back gate contact. The IV curve was measured from single nanowire diodes that were lithographically placed on the substrate. The diodes were contacted by the aluminum contact on the top and via the surface of the substrate. This was done to confirm the contact to the silicon nanowire. As shown in Figure 3, contact was made to the nanowire via the aluminum pad. The resulting single nanowire shows rectification and low recombination current in forward bias. The extrapolated ideality factor is 1.23 below 0.2 V and a series resistance of 384 kohms above 0.6 V. As seen in Figure 3, the plot is not centered about the origin, which is likely due to generation from the light source used on the testing fixture. Contact resistance was measured as 700k ohms before annealing the aluminum and 200k ohms after.

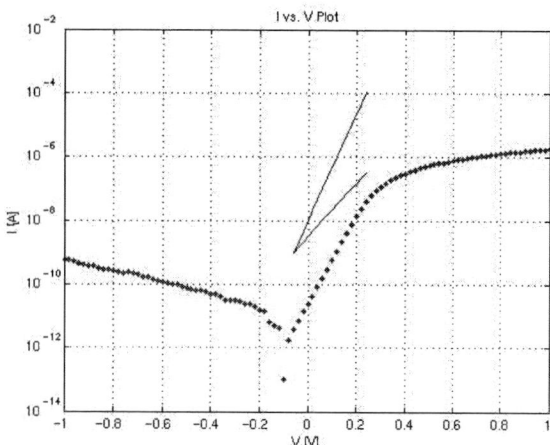

Figure 3: IV curve from lithographically placed single nanowire.

After the back contact was deposited and annealed, more contact measurements were conducted to determine if the rectifying behavior observed was that of the wire or a Schottky contact from the metal probe to the semiconductor surface. IV measurements were performed on the 2 back

contacts and linear behavior was observed. The contact resistance was determined to be approximately 70k ohms. The IV curve depicted in Figure 4 shows the results of measuring from the aluminum back contact, through the substrate and the silicon nanowire to the aluminum pad deposited on the top. As shown here, the nanowire maintains its rectifying behavior.

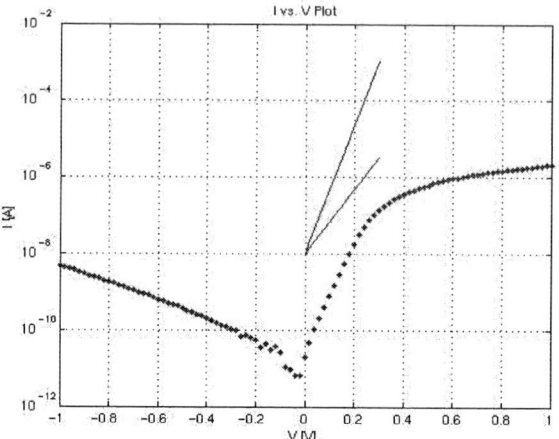

Figure 4: IV curve from silicon nanowire measured from back contact to front contact.

Unfortunately the device illustrated in Figure 3, which consisted of a confirmed single nanowire, failed after the back contact was deposited and annealed. Another nanowire diode structure was tested on the same substrate, which consisted of 1-3 nanowires before the backside aluminum pads were deposited and annealed. As seen in Figure 4, the silicon nanowire diode shows good internal efficiency with an ideality factor of 1.2 below 0.3 V. Above 0.4 V series resistance begins to dominate with a calculated resistance of 350 kohms, attributed to the nanowire.

5. Conlcusion

Single integrated silicon nanowires have been lithographically placed and measured. Rectification has been observed due to the pn junction fabricated within the nanowire. Single silicon nanowires show low recombination currents in the space charge region in forward operation with and ideality factor of 1.2.

6. References

1. Cui, Y., Lieber, C., Functional Nanoscale Electronic Devices Assembled Using Silicon Nanowire Building Blocks, Science, Vol. 291, No. 5505, 2001
2. S.-G. Jun, "Growth and properties of silicon and silicon-germanium nanowires," Ph.D. dissertation, University of Utah, April 2006
3. Tang, Q., Kamins, T., Liu, X., Grupp, D., Harris, J., In Situ p-n Junctions and Gated Devices in Titanium-Silicide Nucleated Si Nanowires, Electrochemical and Solid State Letters, Vol. 8, No. 8, 2005

Design of a MEMS Capacitive Chemical Sensor Based on Polymer Swelling

T. J. Plum, V. Saxena, and J. R. Jessing

ECE Dept., Boise State University, jjessing@boisestate.edu

Abstract— **This paper details the design of a MEMS sensor to detect swelling of polymer films. The sensor is a variable capacitor composed of two electrodes separated by a chemically sensitive polymer. The polymer is chosen such that it absorbs target chemicals (analytes). Upon analyte absorption, the polymer swells, which increases the distance between the two electrodes. This changes the capacitance of the device and can be electrically detected and measured. This paper presents the design of the sensor, the fabrication steps, and polymer selection criteria for the sensor.**

Keywords- MEMS Chemical Sensors, Chemicapacitive Sensors, Chemicapacitors, Polymer-based Sensors.

I. INTRODUCTION

Sensors play a critical role in protecting the public and environment from chemical threats. Micro-electro-mechanical systems (MEMS) are useful as sensors because they are capable of converting different types of signals (e.g., chemical, optical, mechanical) into electrical responses. Small electrical responses can be precisely detected, amplified, and measured using complimentary metal-oxide-semiconductor (CMOS) circuitry. MEMS sensors, especially when integrated with CMOS circuitry, can be highly sensitive (parts per trillion [1]), have fast response times (fractions of a second [2]), and consume little power (as low as nanowatts [3]). These attributes introduce the possibility of in-situ, continuous, and wireless sensing applications. Polymer-based MEMS sensors are popular because polymers react in measurable ways when exposed to small concentrations of target chemicals.

This paper presents the development of a polymer-based MEMS chemicapacitive sensor. The following sections discuss the sensor design, the fabrication steps, and the polymer selection criteria, as well as current progress and future work. While only the sensor is discussed in this paper, the ultimate goal is to integrate the sensor on the back-end of CMOS circuitry for improved sensitivity.

II. DESIGN

The working principle of the sensor begins with a carefully chosen polymer that will absorb target chemicals (analytes). The polymer is sandwiched between two electrodes to form a parallel-plate capacitor, as shown in Fig. 1a. Upon exposure to a target analyte, two significant changes in the polymer's properties, relative to capacitive sensing, occur: (1) the polymer swells [4], and (2) the dielectric permittivity, ε, increases [5], which are demonstrated in Fig. 1b. In general, these changes are proportional to the concentration of the analyte that is present.

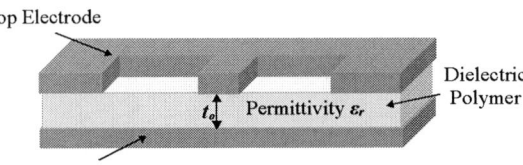

Fig. 1a. Cross-section of a sensing capacitor. The parallel-plate structure is composed of two electrodes separated by a chemically sensitive polymer.

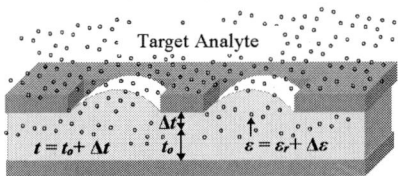

Fig. 1b. A depiction of the polymer dielectric swelling and its permittivity increasing in the presence of a target analyte.

For a parallel-plate capacitor, the metric relating the voltage to current, and thus describing its electrical behavior, is given by:

$$C = \frac{\varepsilon \cdot A}{t} \quad \text{Eq. (1)}$$

where C is the capacitance, ε is the effective dielectric permittivity, A is the overlap area of the electrodes, and t is the distance between the electrodes (this is a simplified equation that neglects fringe capacitances). Equation (1) shows that the capacitance is proportional to the dielectric permittivity and inversely proportional to the distance between the electrodes. Due to the inverse effects that swelling and a change in permittivity have on capacitance, it is possible that the effects could cancel each other. To avoid this, the polymer must be chosen carefully (see Section II-B). Several capacitive sensors using polymer dielectrics have been designed to detect a permittivity change [2,6,7]. This sensor is designed to detect a change in polymer thickness.

A. Sensor Fabrication Steps

To start, a 500nm thermal oxide is grown on a (100) silicon wafer, ensuring sufficient electrical isolation of the sensor from the substrate. Then, 250nm of aluminum is sputtered and patterned, serving as the bottom electrode (Fig. 2a). Next, a polymer is applied using a spin coater to yield a uniform, 1μm thick film. A titanium (Ti) etch mask is sputtered and patterned (using photolithography) on the polymer. The polymer is plasma etched in a barrel asher using O_2 and CF_4 gases (Fig. 2b). The Ti mask remains on the polymer to assist in the final photoresist removal step. Lastly, the top electrode (250nm of aluminum) is sputtered and patterned (Fig. 2c).

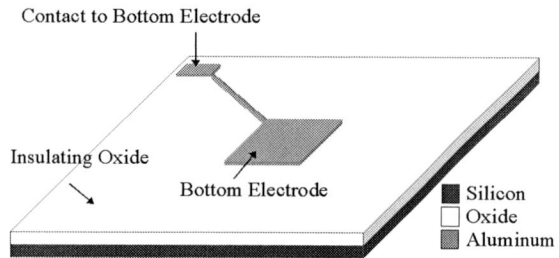

Fig. 2a. After bottom electrode deposition and patterning

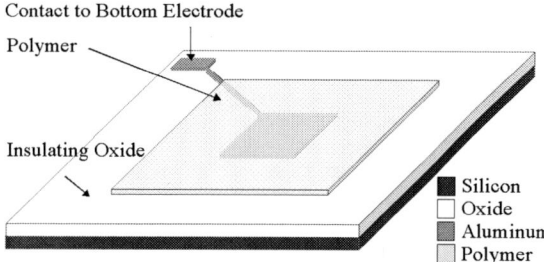

Fig. 2b. After polymer application and patterning

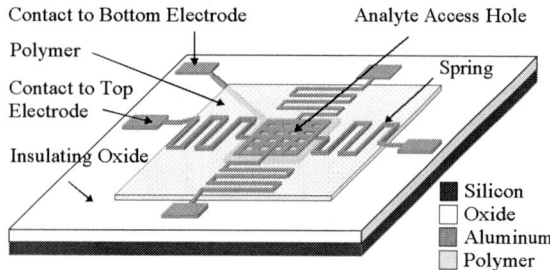

Fig. 2c. After top electrode deposition and patterning

Two important features on the top electrode are shown in Fig. 2c. First, the electrode is porous so that the analyte can access the polymer. This is accomplished by etching an array of holes in the electrode. Secondly, the electrode is supported by planar springs so that it moves when the polymer swells (and thus does not significantly impede swelling). Sensors are being fabricated that vary the size and density of access holes as well as the length and width of the springs. The dimensions of the electrodes range from 450µm by 450µm to 750µm by 750µm and the baseline capacitances range from approximately 1pF to 10pF.

B. Polymer Selection Criteria

Polymer selection is critical in being able to detect swelling. Most importantly, the polymer must absorb the analyte. Methods to predict this absorption, such as the Hansen solubility parameters or the linear solvation energy relationships, have been reported [8,9]. Next, the polymer swelling must be reversible, which typically requires that only weak bonds form between the polymer and analyte [9]. Thirdly, the analyte permittivity should be as small as possible relative to the polymer permittivity. This will help minimize the permittivity change so it does not negate the effects of polymer swelling. Lastly, because of the heat associated with some of the fabrication processes, the polymer should have a glass transition or melting temperature greater than 80°C.

The polymer that is currently being experimented with is poly(ethylene-vinyl acetate) (PEVA). PEVA has shown swelling behavior when exposed to toluene and benzene. Adsorption of chemical warfare agent simulants dimethyl methylphosphonate (DMMP) and cholorethyl ether (CEE) have also been reported [2].

III. PROGRESS AND FUTURE WORK

Sensors are currently being fabricated. When completed, they will be bonded to dual inline packages (DIP) and tested in an enclosed chamber. The gas content and temperature in the chamber will be controllable. Eventually the sensors will be integrated with CMOS circuitry and measured via differential capacitance (comparing sensor with a reference capacitor).

IV. CONCLUSION

A method of detecting polymer swelling using a MEMS capacitor has been proposed. The design and fabrication steps have been presented. First-run sensors are currently being fabricated. Future work includes significant device characterization and sensor integration with CMOS circuitry for improved sensitivity.

Support of this project by EPA Contract no. X-97031101-0 is gratefully acknowledged.

REFERENCES

[1] R.A. McGill, R. Chung, D.B. Chrisey, P.C. Dorsey, P. Matthews, A. Pique, T. Mlsna, and J.L. Stepnowski, "Performance Optimization of Surface Acoustic Wave Chemical Sensors," IEEE Transactions on Ultrasonics, Ferroelectrics, and Frequency Control, vol. 45, no. 5, Sep. 1998, 1370-1380.

[2] S.V. Patel, T.E. Mlsna, B. Fruhberger, E. Klaassen, A. Cemalovic, and D. R. Baselt, "Chemicapacitive microsensors for volatile organic compound detection," Sensors and Actuators, B. Chemical, vol. 96, no. 3, pp. 541-553, Dec 1, 2003.

[3] J.D. Adams, G. Parrott, C. Bauer, T. Sant, L. Manning, M. Jones, and B. Rogers, "Nanowatt chemical vapor detection with a self-sensing, piezoelectric microcantilever array," Applied Physics Letters, vol. 83, no. 16, Oct. 2003, pp. 3428-3430.

[4] S. Chatzandroulis, D. Goustouridis, I. Raptis, "Characterization of polymer films for use in bimorph chemical sensors," Journal of Physics: Conference Series, vol. 10, 2005, pp. 297-300.

[5] A. Koll, A. Kummer, O. Brand, and H. Baltes, "Discrimination of volatile organic compounds using CMOS capacitive chemical microsensors with thickness adjusted polymer coating," Proceedings of SPIE - The International Society for Optical Engineering, vol. 3673, pp. 308-317, March, 1999.

[6] R. Ishihara and S. Matsubara, "Capacitive Type Gas Sensors," Journal of Electroceramics, vol. 2, no. 4, 1998, pp.215-228.

[7] H. Shibata, M. Ito, M. Asakursa, and K. Watanabe, "A Digital Hygrometer Using a Polyimide Film Relative Humidity Sensor," IEEE Transactions on Instrumentation and Measurement, vol. 45, no. 2. Ap. 1996, pp. 564-569.

[8] C.M. Hansen, Hansen Solubility Parameters, A User's Handbook, CRC Press, Boca Raton, FL, 1983.

[9] J.W. Grate and M.H. Abraham, "Solubility interactions and the design of chemically selective sorbent coatings for chemical sensors and arrays," Sensors and Actuators: B, vol. 3, 1991, pp. 85-111.

Introduction to Modeling an Imaging System
With Human Detection of Artifacts

[*] Vinesh Sukumar, [*]Doug Warner, [*]Patrick Doherty
[**]Herbert Hess, [**]Ken Noren, [***] Steve Krone

[*] *Micron Technology, Inc., 8000 S. Federal Way, Boise, Idaho 83707-0006*
[**]*Microelectronics Research and Communications Institute, University of Idaho, Moscow, Idaho, U.S.A.*
[***]*Department of Mathematics, University of Idaho, Moscow, Idaho, U.S.A*

Abstract: This paper introduces a mathematical model to help identify defects for any imaging system. These defects become detectable when the combined information presented by an imaging system exceeds the perception threshold of a human visual system. These models are being developed under certain simplifying assumptions which are presented in the course of the paper.

Keywords: CMOS Image Sensor, Defect Modeling and Noise

I. INTRODUCTION

Over the years, their has been an increased need to come up with an objective way to measure perceived image quality. This is limited not only by the physical parameters of the image forming system, like resolution and contrast, but also on the impression of the image formed by the eye of the observer which has is own limitations. This is gaining significant importance for engineers working in CMOS Imaging industry. Design for test/pixel defect correction becomes easier to implement as the limitations for perceived quality of an image are better understood with an image that can be projected without a real sensor in place.

The scope of this paper is limited to introducing a unique defect detection model. This model includes human detection and image processing in terms of contrast sensitivity defined in a spatio-temporal domain, optical and digital processing system mentioned in terms of MTF. These models will be based on weighted combination of the physical parameters of the image forming system and psychophysical parameters of the human visual system and presented in mathematical expressions which can be easily used in practical applications. Presenting mathematical equations are beyond the scope of this paper. This model will be helpful to calculate the critical flicker frequency for Image sensors, which can further be used for better technical design of these systems. These models will be restricted to foveal vision wherein the maximum efficiency is observed. The image quality metrics with emphasis on practical measurements will be presented in the conference. The image quality measurements need to be obtained by a panel of observers that look at images under carefully standardized viewing conditions to gauge

the performance of the model. The goal of this work is to quantify quality for a batch of given sensors or to simulate noise in a digital camera simulator.

II. MODELING AND ESTIMATION

Assuming constant illumination, an Imaging system is modeled as presented in Fig.1. Certain simplifying assumptions were made to keep the model simple and as accurate as possible. Information presented from an optical system for a target object is seen as combination of optics and stand alone sensor. Our first assumption that loss of information can be modeled as a combination of the optical properties of the optics and a noise model for the sensor. This unified model is represented as MTF of the optical system. With increased emphasis on image quality standards, several low-power, highly integrated, hardware and programmable camera signal processing platforms have been developed. These on-chip processing engines can lower presented information by the sensor which can also be modeled as MTF of the image processing system. The human visual system plays an important role in making clear distinctions. Objects can generally be distinguished, if the difference in luminance or color is large. In practice the relative difference in luminance plays an important role in decision making process. This can be expressed in terms of contrast ratio, or the difference between two luminance values divided by the sum of them, which can also be expressed as contrast. The reciprocal of the minimum contrast required for detection is termed contrast sensitivity. These models for a human detection system are evaluated in terms of modulation threshold, which is defined as a fixed threshold below which a luminance variation is not observed and above which luminance variation is always observed. The

---

*Corresponding Author. Tel + 1-208-368-1654.
E-mail address: vsukumar@micron.com.

These defects can be detected in the final perceived image only if
(Image produced by the Optical System) * (Image processing done by on-board chip electronics) > Detection threshold of a human visual system. This is under the assumption that there is no loss of information along the entire data path transmission

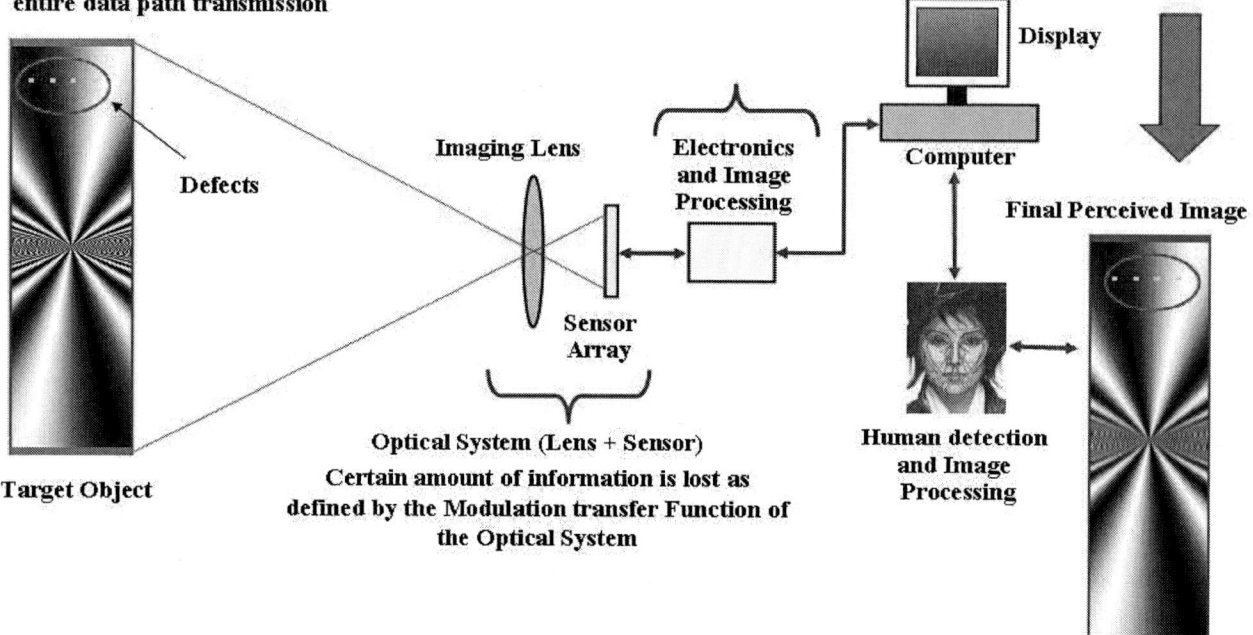

Figure: 1 Block diagram representation of an Imaging System [8]

final perceived image standards can be predicted based on mathematical expressions for MTF of an Optical System, Image Processing System and Contrast sensitivity measurements of the Human Visual System. Image quality metrics are usually generic measurements. This would translate to being independent from actual picture content. Several image quality metrics have been developed over the past few decades. The functional parameters of these metrics are under study at this time. Research has shown that metrics based of mathematical expressions lack good correlation with subjectively perceived image quality [7]. To better quantify validity of the developed model, a panel of observers will be used for an objective measure to gauge the performance of the model.

III. CONCLUSION

An attempt is made to present a model to quantify captured image from an imaging system and to measure perceived image quality. These models will be based of weighted combination presented in mathematical expressions. They will be derivatives based of the physical parameters of the image forming system and psychophysical parameters of the human visual interface which will be presented in the conference.

A number of issues remain to be investigated in terms of understanding loss of resolution based of on-chip image processing algorithms. More data and analysis has to be completed in understanding image quality metrics and standards of measurement.

Finally research has to be done to understand the limitations of the model with continuous efforts to extend the scope of the model for a more broadened usage.

REFERENCES

[1] B. Fowler, J. Balicki, D. How, S. Mims, J. Canfield and M. Godfrey, "An Ultra Low Noise High Speed CMOS Scientific Line scan Sensor for Scientific and Industrial Applications," *2003 IEEE Workshop on Charge-Coupled Devices and Advanced Image Sensors*, May 15-17, 2003, Elmau, Germany.

[2] B. Fowler, A. Krymski, N. Khalliulin, H. Rhodes, "A 2 e- Noise 1.3 Megapixel CMOS Sensor, *2003 IEEE Workshop on Charge-Coupled Devices and Advanced Image Sensors*, May 15-17, 2003 , Elmau, Germany.

[3] D.Yang, B.Fowler, A.El Gamal, H.Min, M.Beiley and K.Cham, "Test Structures for Characterization and Comparative Analysis of CMOS Image Sensors, "in *Proceedings for SPIE*, October 1996.

[4] Hank Hogan, *"Image is Everything"*, pp.82, Photonics Spectra, Dec 1998.

[5] Kalwant Singh, "Noise Analysis of a fully Integrated CMOS Image Sensor", in *Proceedings for SPIE*, Vol. 3650, p. 44-51. Mar 1999.

[6] A.El Gamal, B.Fowler H.Min, X.Lin, "Modeling and Estimation of FPN components in CMOS Image Sensors, "in *Proceedings for SPIE*, Vol. 3301, Jan 1998.

[7] Barten P.G.J, "Spatio-temporal model for the contrast sensitivity of the human eye and its temporal aspects, "*Human Vision, Visual Processing and Digital Display IV, Proc. SPIE*, Feb 1993.

[8] Randy Linebarger, "MTF Characterization, "*Imaging Product Characterization Seminar*, Micron Technology, Inc. Feb 2005

[9] Gennady Agranov, "Basic Opto-Electrical Characterization Methodology for Sensor Core, "*Imaging Product Characterization Seminar*, Micron Technology, Inc. Feb 2005

Comparative study of TaN-TiN and TiN gate stacks for thermally stable PFETs

Nirmal Ramaswamy, Allen Mcteer, Venkat Ananthan, Nanda Palaniappan, Tim Owens,
Sanh Tang, Ravi Iyer, Shixin Wang, Chandra Mouli.

R&D Process Development, Micron Technology, Inc. Boise, ID

(Phone) 208 368 1582, (fax) 208 368 2548, (e-mail) dramaswamy@micron.com

Abstract—Thermally stable TiN and TaN-TiN laminate metal gates for PFETs are demonstrated using a conventional CMOS process flow with SiON gate dielectric. TaN-TiN laminate gates show enhanced drives (I_D), higher transconductance (G_M), higher mobility (μ_{EFF}), and reduced off current (I_{OFF}) characteristics compared to TiN gates. The optimum thickness of TaN in the laminate stack is discussed. The as-deposited work function of the TaN-TiN laminate gate stack and TiN was found to be ~ 5.0eV.

Index Terms—CMOS, TaN, Laminate metal gate, Work function, Metal nitride.

Introduction

With aggressive scaling of CMOS devices, polysilicon gate material eventually needs to be replaced with an appropriate metal gate stack. Currently, heavily doped p-type and n-type polysilicon is used to design PMOS and NMOS devices. However, the solid solubility and activation of dopants are limited, resulting in poly-depletion. To achieve optimum CMOS device performance, metal gates with work functions (WF) close to the Si band edges are preferred. However, identifying and integrating two thermally stable metal gates that possess the desired CMOS device WF is challenging. Hence, researchers are investigating novel methods to enable dual work functions by manipulating the metal electrode/gate oxide interface to simplify integration. Alloying [Ru-Ta[1], HfN[2], silicides[3]], implantation [Mo-N implant[4]] and multilayer[5] stacks have been studied by various groups. Bilayer and multilayer stacks have been recent topics of research for improving thermal stability and optimizing metal gate electrode WF[1-5]. In this letter we have investigated and compared device and material characteristics of TaN-TiN laminate gate stacks to TiN gate stacks with respect to high WF and thermal stability characteristics.

Device Fabrication

Material analysis of TaN and TiN films: CVD (Chemical Vapor Deposition) TiN, ALD (Atomic Layer Deposition) TiN and ALD TaN have been investigated by various groups as candidates for metal gate electrodes[6, 7]. Excellent thickness control, good step coverage and high thermal stability make these materials good candidates for metal gate electrodes for planar and nonplanar devices. In our work, ALD TaN was deposited by using a metal-organic precursor and NH_3. CVD TiN was deposited using $TiCl_4$ and NH_3. X-ray reflectivity (XRR) and X-ray diffraction (XRD) analyses were performed on 100Å as-deposited and heat-treated blanket films to characterize material properties. As observed from the XRD patterns (Figs. 1-2), TiN films are crystalline as deposited and grain growth post-heat treatment is minimal. TaN films are nanocrystalline as deposited and exhibit a high degree of crystallinity post 1,000°C 20s anneal. The maximum observed density change in TaN was approximately 30 percent (9.16–12.13 g/cc) and 10 percent (4.64–5.08 g/cc) for TiN (Figs. 3-

4). MOS capacitors were fabricated on SiON gate dielectric with TiN and TaN/TiN laminate gates. Impact of TaN film thickness on the CV curve of laminate was studied. WF was obtained from V_{FB} (Flat-band) vs EOT (Effective Oxide Thickness) plots. The as-deposited WF was calculated to be ~5.0eV for both stacks (Fig. 5). Using these gate stacks, PFETs were fabricated with 24Å EOT SiON gate oxide. Process flow is schematically described in Table 1. Thicker TaN films (>30Å) show a shift in the CV curve post-1,000°C, 20s rapid thermal anneal (RTA) (Fig.6).

Results and Discussion

As observed from the extracted WF values, both devices have similar V_T of approximately –0.7V (–0.7V for TaN-TiN; –0.73V for TiN gates). TaN-TiN laminate gates show 45 percent higher drive current (at drain to source voltage-V_{DS} = 1.65, gate voltage-V_G = –2.0V) and 13X lower off-state leakage (Fig. 8). This result is consistent with the higher peak transconductances (Fig. 9), higher effective mobility (Fig. 11) and lower subthreshold slopes observed with TaN/TiN gates. TaN/TiN gates also exhibit an order of magnitude lower gate induced drain leakage (GIDL) than TiN devices (at gate to drain voltage V_{GD} = 3V) and 30X lower gate leakage current (at V_G = –3V, Fig. 8). Fig. 7 shows the high-resolution transmission electron micrograph (HRTEM) cross-section of TaN-TiN and TiN stacks on gate oxide post-1,000°C, 20s RTA. The TaN/SiON interface is much more distinct and defined than the TiN/SiON interface. This improved interface is probably the cause for the observed better performance of the laminate electrode.

Conclusions

High-WF, thermally stable PFETs have been demonstrated with TaN and TaN-TiN laminate gates. The optimum thickness for maintaining high-gate stack WF has been discussed. Microstructural changes in thicker TaN films (>30Å) in the laminate stack could lead to shifts in the CV curve post-heat treatment. The improved properties of TaN-TiN laminate devices are believed to stem from the much improved metal-to-gate oxide interface in the laminate gates.

References

[1] V. Misra, et al.., IEEE Electron Dev. Lett., Vol. 23, 354-356 (2002).
[2] W.P.Bai et al., IEEE Electron Dev. Lett, Vol. 26 No 4, 231-233 (2005)
[3] C. Cabral, Jr. et al., Symp. VLSI Tech., 184 (2004).
[4] R. Lin et al., IEEE Electron Dev. Lett., Vol. 23, pp. 49-51 (2002)
[5] Ching-Huang Lu, et al. ,IEEE Electron Dev. Lett, Vol. 26,445-447 (2005)
[6] D.G. Park, et al., J. Electrochem. Soc., 148, 9, F189-F193, 2001.
[7] J. Westlinder, et al., IEEE Elect Dev Lett, v24, n9, p550-552, 2003.

1	STI/CMOS Channel I/I	6	Spacer Formation and S/D I/I
2	SiON (28A)	7	Activation Anneal (1000C,5s)
3	Electrodes : ALD TaN-TiN or CVD TiN	8	Source Drain Contact Metallization
4	Gate patterning	9	ILD
5	LDD/Halo I/I	10	Back end Metallization / Patterning

Table 1. Process flow of conventional MOSFET.

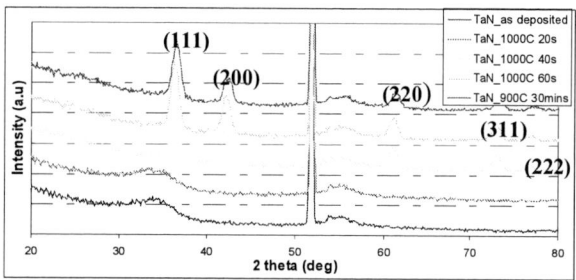

Fig. 1. XRD pattern showing TaN microstructure evolution with heat treatment.

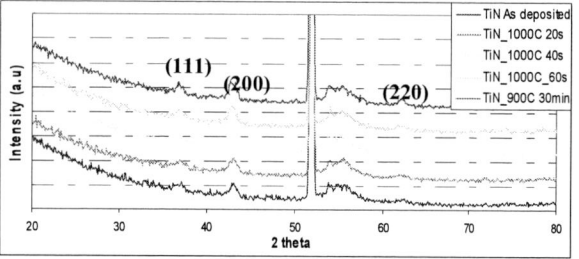

Fig. 2. XRD pattern showing TiN microstructure evolution with heat treatment.

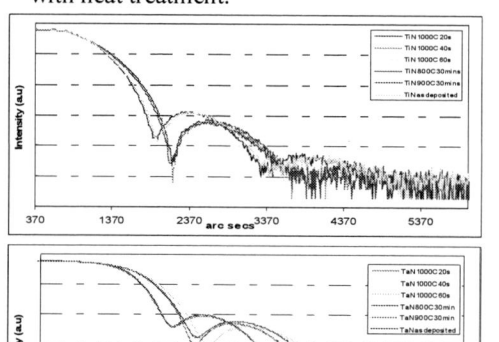

Fig. 3. XRR spectra for TiN.

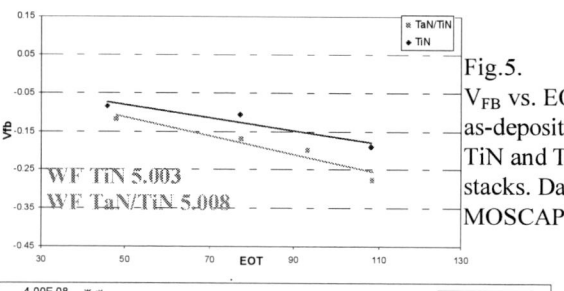

Fig. 4. XRR spectra for TaN.

Fig.5. V_{FB} vs. EOT plots for as-deposited TaN-TiN and TiN gate stacks. Data from MOSCAPs.

Fig.6. CV curves for various thicknesses of TaN in TaN-TiN laminate stack from MOSCAPs, pre (inset) and post HT.

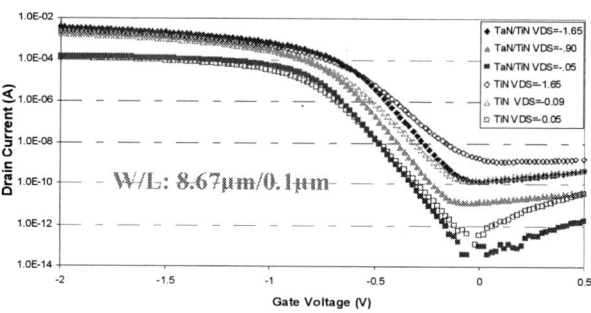

Fig.7. TEM Micro graph of SiON/Metal gate interface post 1,000°C, 20s RTP.

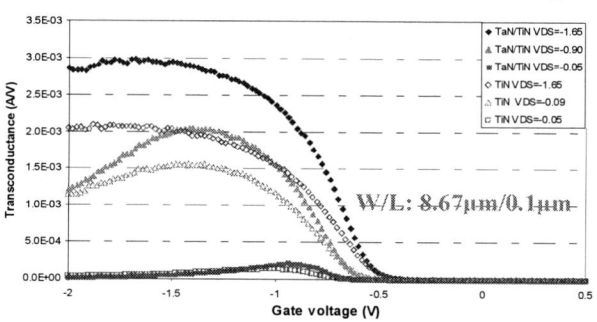

Fig. 8. I_D-V_G characteristics: lower I_{OFF} with TaN-TiN gate.

Fig. 9. G_M-V_G characteristics of PFETs. The peak G_M of TaN-TiN is 40% higher than that of TiN gate(V_{DS}=-1.65V).

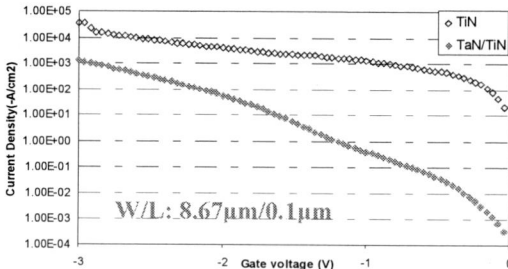

Fig. 10. J-V characteristics of PFETs: Lower gate current for TaN-TiN gate.

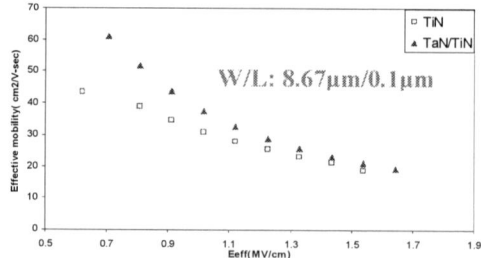

Fig. 11.Effective Hole Mobility Versus E field: higher μ_{EFF} with TaN-TiN gate.

54
1-4244-0374-X/06/$20.00 © 2006 IEEE

Dependence of Si₃N₄ Film Properties on Precursor Chemistry

Fernando Gonzalez, Senior Member; Shyam Surthi, Member; Parag Banerjee, Member

Micron Technology, Incorporated

Boise, ID, USA

Abstract— **Silicon nitride films are used as oxidation barriers, mobile ion barriers, hard masks and capacitor dielectrics in integrated circuit manufacturing. Properties of silicon nitride layers are characterized relative to physical, optical, chemical and electrical aspects that pertain to specific applications. The process integration of silicon nitride within a device structure depends upon uniformity and controllability of the deposition conditions. Precursors used in the silicon nitride deposition reaction will affect stoichiometry and reaction rates. The Trisilylamine (TSA) precursor is compared to standard Dichlorosilane (DCS) in terms of performance parameters.**

Keywords- Silicon Nitride, Dielectric

I. INTRODUCTION

Silicon nitride (Si_3N_4) has been used since the late 1960s; Dr. Kooi [1] used its oxidation retardation properties to block high temperature and long-processing-time field oxidation. The main issue found with Si_3N_4 was its tensile stress, which induced defects under the field edge "bird's beak." The defects would degrade diode leakage in the circuits, especially in DRAM memory cells. The Si_3N_4 film was also used as a hard mask, to define the gate stack and to protect the polycide from oxidation. Stresses in Si_3N_4 films provide another disadvantage by affecting gate-edge dielectric breakdown. Si_3N_4 dielectric has a higher k value (permittivity) than silicon dioxide. Consequently, Si_3N_4 was chosen to replace silicon dioxide as a capacitor dielectric material. The precursor, TSA, has been considered for a low-temperature gate stack process.

II. METHODS OF DEPOSITION

A. Reactants used in silicon nitride deposition

- *DCS chemical reaction and symbols*
  Reaction 1, $\Delta = 600°C–800°C$

$$3Si_2Cl_2H_2 + 8NH_3 \rightarrow 2Si_3N_4(s) + 3Cl_2\,(g) + 15H_2\,(g)$$

DCS is a chlorinated silane gas and the source of silicon in Reaction 1. DCS gas pyrolyzes above 550°C into an amorphous silicon film. The DCS-based with NH3 reaction will leave approximately 1 percent of chlorine throughout the bulk of the Si_3N_4 film within XPS detection limits.

- *TSA chemical reaction and symbols*
  Reaction 2, $\Delta = 500°C–800°C$

$$3(SiH_3)\,N + 3NH_3 \rightarrow Si_3N_4(s) + 9H_2\,(g)$$

TSA is a precursor used for low-temperature Si_3N_4 films [2]. Examining the free energy change associated with reactions (1) and (2) shows that, at the same temperature, TSA-based Si_3N_4 has a lower ΔG as compared to DCS-based

Si_3N_4, indicating more thermodynamic 'drive' for Si_3N_4 formation. This calculation is shown in Figure 1 [3]. TSA bonds three silylamines to one nitrogen atom and is the silicon source for Reaction 2. There is no chlorine in this reaction. There is an advantage in particle reduction as measured with laser light particle detectors since there is no NH_4Cl formed (fig. 2).

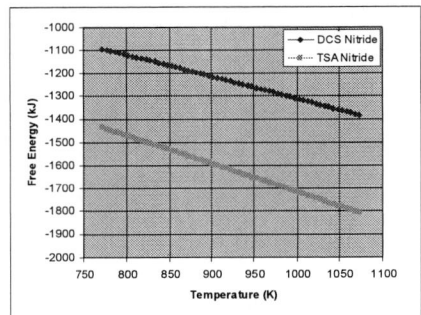

Figure 1: ΔG (free energy change) for reactions 1 and 2, indicating more thermodynamic 'drive' for TSA vs. DCS.

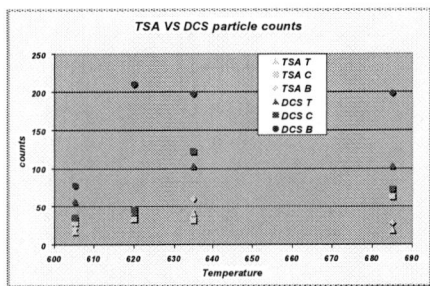

Figure 2: Particle formation between TSA and DCS.

III. EXPERIMENT AND RESULTS

A. X-ray Photoelectron Spectroscopy (XPS)/ index of refraction (n)

Chemical bonding energies for both TSA and DCS films analyzed with XPS (fig. 3) and ellipsometry indicate that the stoichiometry and refractive index variations are comparable. Even though precursor thermodyamic properties (Cp) are different, the reactions give stoichiometric silicon nitride (n=2).

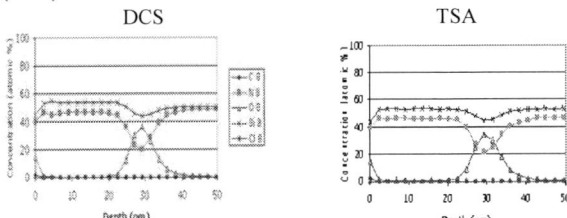

Figure 3: Oxidized Si-N measured with XPS data, comparing DCS and TSA. Samples were oxidized with O/H radicals at 800°C.

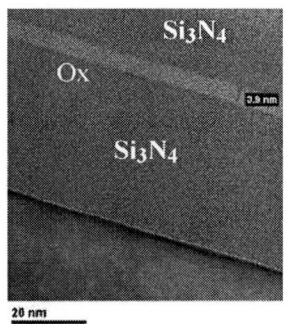 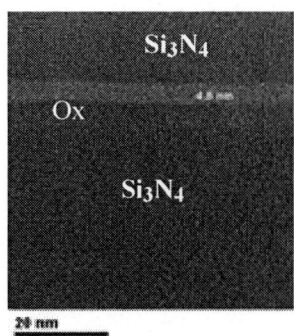

TEM Picture 1: DCS (left) vs. TSA (right).

The TEM thickness reading of the oxidation on the silicon nitride film (TEM Picture 1) was compared to the oxide thickness on a separate bare silicon wafer measured by ellipsometry. The ratio of the oxynitride and the thermal oxide is approximately the same ratio for both types of film (TSA=42A, DCS= 39A for a 70A bare silicon thermal oxide).

Using the XPS technique, the percent of chlorine was determined to be below the level of Cl detectability for TSA, however, the Cl level was approximately 0.8 percent Cl for the DCS film. On the other hand, the index of refraction was much higher when the ammonia to TSA precursor was low. The 200sccm of NH3 to 20sccm of TSA gave 2.15 refractory index versus the 480 to 20 sccm gave 2.0 refractory index.

B. Thickness and deposition rates

TSA-Si₃N₄ deposition does not have a good uniformity performance compared to DCS-Si₃N₄ deposition as shown in TSA data in figure 4. Spacing wafers on the boat every other slot (2X) or every fourth slot (4X) helps uniformity, but it limits the number of wafers that can be processed during a run. The boat rotation did not help with the TSA across the wafer uniformity. The uniformity was limited by the mass flow dynamics shown in figure 5.

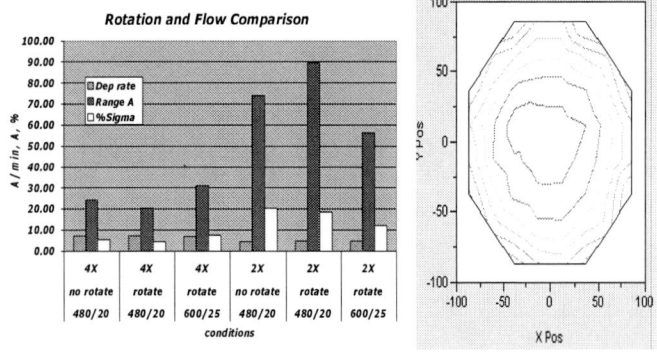

Figure 4: Deposition rate, uniformity (with boat rotation), wafer slot spacing, and NH₃/TSA flow ratio. The TSA contour map shows across wafer variation. (600C)

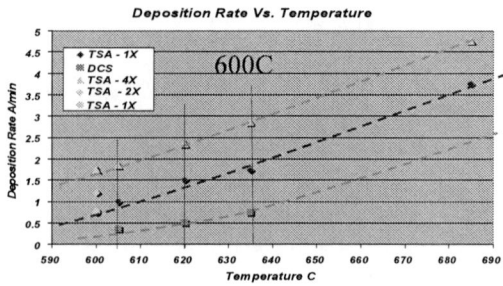

Figure 5: Deposition rates of different precursors vs. wafer slot spacing.

D. Electrical Data

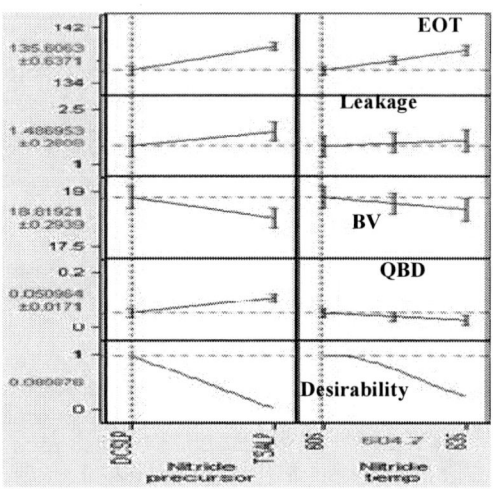

Figure 6: ONO comparison in EOT thickness, leakage, BV, and QBD [4]

In figure 6, the electrical comparison shows that TSA-Si₃N₄ performing with higher leakage, lower breakdown voltage, and higher electrical Equivalent Oxide Thickness (EOT) than DCS- Si₃N₄. The electrical differences may relate to the TSA-Si₃N₄ nitride oxidation properties and its index of refraction.

IV. CONCLUSIONS

TSA is an attractive, low-temperature deposition precursor. It has a better deposition rate at lower temperatures than the DCS precursor. However, TSA silicon nitride uniformity is a major hurdle to overcome for practical applications. Also, DCS performs better electrically in an ONO application than TSA. However, the 600°C deposition temperature electrically performs better than the higher temperatures for both precursors. In this study we have shown low temperature deposition capability (600C) with TSA and DCS precursors.

REFERENCES

[1] Appels,J.A.,Kooi,E.,Paffen,M.M.,Schatorjé,J.J.H.andVerkuylen, W.H.C.G. Philips Research Reports Vol. 25. 1970, pp, 118-132.
[2] McKean DC, Torto I. *Spectrochim Acta A Mol Biomol Spectrosc.* 2001 Aug; 57(9):1725–38
[3] Stock and Somieski, Berichte 740, 750 (1921)
[4] F.Gonzalez and A. Howard, Six Sigma Black Belt Project

Photo Sensitivities in a 0.35μm 18V PDMOS Technology

Brett Williams, Mike Thomason, Chuck Belisle
AMI Semiconductor, Inc.
Pocatello, Idaho, USA
Brett_Williams@amis.com

Abstract— **The I3T25 technology being developed at AMI Semiconductor, Inc. uses lateral extended-drain MOS transistors (DMOS) [1] in a 0.35μm base technology. These devices are very sensitive to the well and field implant critical dimensions (CDs) and the layer-to-layer alignment (overlay). This sensitivity is much greater than the standard CMOS devices. The photoresist used on the P-channel field implant (PFLD) mask has a strong dependence on reticle transmission (RT) and has caused variability in the P-channel DMOS performance.**

Keywords-component; Extended Drain MOSFET, Smart Power Ics, RESURF, photomasks

I. INTRODUCTION

Medium-Voltage (>16V) lateral extended-drain MOS transistors (DMOS) are currently of great interest for automotive, industrial, and medical applications (e.g., pacemakers, neurostimulators, etc.), requiring very high reliability standards. The I3T25 technology was developed at AMI Semiconductor located in Pocatello, Idaho to address this market. The technology is built on a 0.35μm technology base and incorporates specific isolation rules to allow for high voltages. The spacing of key dimensions such as implanted Wells and Fields are critical in producing a reliable, manufacturable device.

A new testchip (19427) was developed using a light-field reticle for the P-channel field (PFLD) implant. This implant is used as a Reduced Surface Field (RESURF) implant [2-3] for the 18V P-channel DMOS structure (PDMOS18) as shown in Fig. 1. The previous testchip (19234) used a dark-field reticle for this implant and incorporated the CMOS PWELL implant in the same process step. The breakdown voltages (BVDSS) were 7-8 volts lower, the on-resistance (RON) was 50 ohms lower, and the threshold voltage (VT_GM) was 50mV higher on the same PDMOS18 device versus the previous testchip (see Table 1).

TABLE I. COMPARISON OF ELECTRICAL RESULTS OF THE SAME PDMOS18 STRUCTURES ON THE 19234 AND 19427 TESTCHIPS

Product	Structure	BVDSS	RON	VT_GM
19427	PDMOS18	-24.388V	481.211Ω	-0.674V
19234	PDMOS18	-31.759V	527.067Ω	-0.625V

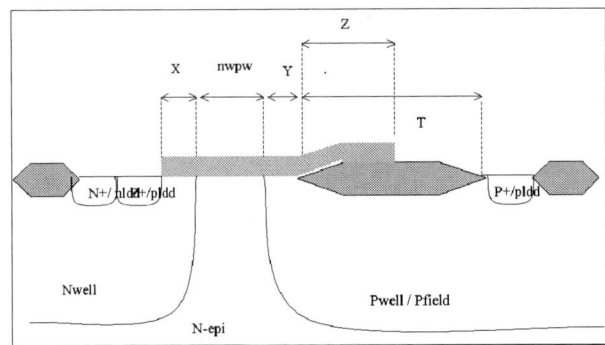

Figure 1. - I3T25 PDMOS18 Structure Cross-section

A reduction in the exposure energy at the PFLD mask is a temporary fix to this issue. However, it is desired that a more robust photoresist be qualified at the PFLD mask step.

II. INVESTIGATION AND THEORIES

The known differences between the processing of the two testchips were the first to be investigated. Since each testchip needed to use a different process flow, these flows were reviewed very carefully. All was in order.

The implant doses were checked to make sure that the new 19427 lots didn't receive a different amount of dopant at PFLD. The photoresist thickness of each lot was confirmed just in case extra dopant was bleeding through a thin photoresist at PTUB. SIMS were done in an area of PFLD under FOX comparing the 19427 with the 19234 testchips to determine if there was a possible N-EPI or PFLD doping difference. The results of this SIMS can be found in Fig. 2. There is a slightly different dose level at the very surface that has not been fully explained.

The Etest data was also verified to make sure that this difference was not some testing issue. Additional structures were also reviewed. It was found that on a 19427 PDMOS18 structure where the Nwell was intentionally pulled back to the left (see Fig. 1), the BVDSS and RON values matched the 19234 standard device much better. This was an important clue to finding the root cause of this issue.

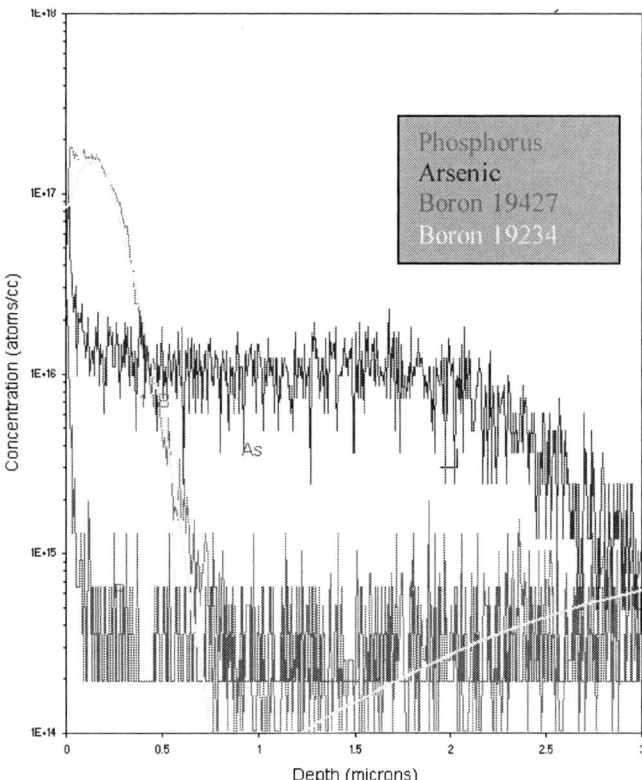

Figure 2. SIMS comparison in PFLD area

TABLE II. COMPARISON OF PFLD OPENINGS IN THE TWO TESTCHIPS FROM SEM VS. DRAWN MEBES DATA.

Product	SEM1	SEM2	MEBES	Difference
19427	3.06μm	3.13μm	2.70μm	~0.4μm
19234	5.61μm	5.73μm	5.70μm	~0.0μm

TABLE III. DESCRIPTION OF VERIFICATION EXPERIMENT

					WAFER NUMBER			
OP Name	Photoresist	Reticle	FEM	CD Target	1	2	3	Comment
TUBMSK	PRG 6	Rev A	Yes	2.05um	X			230mJ w/ 35mJ steps
	PRG 6	Rev A	No	2.05um		X	X	
PTUBMSK	PRG 6	Rev A	Yes	2.06um		X		230mJ w/ 35mJ steps
	PRG 6	Rev A	No	2.06um	X		X	
OP Name	Photoresist	Reticle	FEM	CD Target	1	2	3	Comment

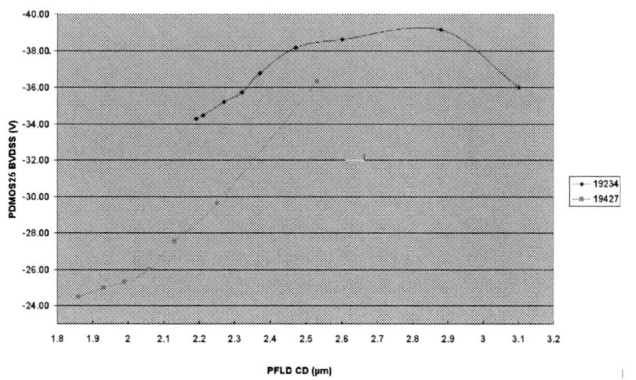

Figure 3. PDMOS18 BVDSS results from 19234 and 19427

The reticle data (MEBES) from the two testchips was also reviewed and compared to see if there was a difference in the fracture that may have caused such a shift. Although the structures were not identical since they were created at different times by hand, the critical measurements were the same. The PFLD opening was measured on MEBES for both testchips on the standard devices. Then shortloop wafers were processed for each of the masking steps in question (PFLD, NTUB, PTUB) for each testchip. These wafers were masked only, leaving the photoresist. The wafers were then submitted for SEM. Once the data was reviewed, it was easy to see that there was a significant difference in the PFLD MEBES to photoresist SEM offset as seen in Table 2.

III. VERIFICATION EXPERIMENT

An experiment using the two testchips is described in Table 3. As a note, the CDs referred to in this section are photoresist bars in the scribe line. Therefore, if the CD is larger, the photoresist opening for the implant is smaller. For example, since the PFLD photoresist opening on the 19427 testchip is too large compared to MEBES data (according to Table 2) it will be necessary to increase the PFLD CD measurement to match the PDMOS18 results from the 19234 testchip.

The most significant results of the experiment are that the BVDSS is not as strongly affected by the PFLD CD on the 19234 as the 19427 testchip (see Fig. 3). However, if the PFLD reticle were biased down on the 19427 reticle by ~0.4μm, the plots of the 19234 and 19427 results in Fig. 3 would likely line up. Also, the 19234 data shows that if the PFLD CD is enlarged too much there will eventually be a reduction in BVDSS. To obtain the highest photoresist CDs, the exposure energy is significantly reduced from 230mJ to just 90mJ. This causes scumming of the photoresist in the openings. The scumming is likely the source of this decrease in BVDSS at the lowest exposure energy.

Further work on this issue involves biasing down the PFLD reticle data on the 19427 and repeating the experiment. Also, a photoresist is being analyzed that will be less sensitive to RT changes.

REFERENCES

[1] A. Moscatelli, A. Merlini, G. Croce, P. Galbiati, C. Contiero, "LDMOS Implementation in a 0.35μm BCD technology (BCD6)", ISPSD2000, p. 323, May 2000.

[2] A. W. Ludikhuize, "Review of RESURF Technology", ISPSD2000, p. 11, May 2000.

[3] J. A. Appels, H. M. J. Vaes, "High Voltage Thin Layer Devices (RESURF DEVICES)", IEDM, p. 238, 1979.

Silicon Solar Cells Using Backside Contacts With Through-Wafer Interconnects

Aaron Erbe and A.J. Moll
Materials Science and Engineering
Boise State University

Abstract – **In this paper, a viable alternative design for silicon solar cells using backside contacts with through-wafer interconnects (TWI) is discussed. TWI technology allows conductive paths to be created without sacrificing top surface area. Using this technology, we will be able to provide a connection from the top n-type material to the cathode at the rear of the solar cell. The potential advantages and the proposed design are also discussed.**

I. INTRODUCTION

Photovoltaic Cells can be a viable and competitive alternative energy source with improvements in efficiency made by novel design methods.

A. Traditional Design Issues
In traditional solar cell designs (Fig. 1), light must pass through the n-type material with the cathode designed as a metal contact grid. This contact grid compromises useable surface area and results in a shadow effect which lowers the conversion efficiency of the solar cell.

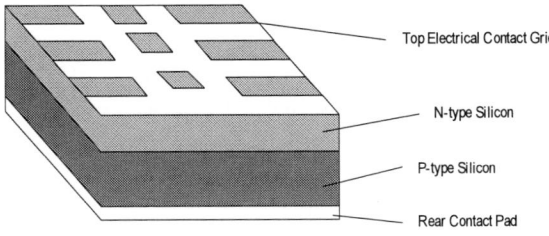

Figure 1 Typical solar cell design

B. Potential Benefits to Backside Contact Design.
Backside Contact (BC) design is one method that has been under development and shown to significantly increase the efficiency of solar cells. The BC design attempts to create interconnection from the n-type material to the cathode without compromising the top surface area. Using Through-Wafer Interconnect (TWI) technology,

further improvements to BC designs can be made allowing for higher efficiency rates.

The advantages to using TWIs would be:
- Increased surface area allowing more absorption of photons
- Eliminated "shadowing" effect caused by traditional contact grid pattern
- Potential for a higher density cell design by removing the need for interconnecting Bus Bars.
- TWI design may also help transfer heat away from the junction, lowering the operating temperature and increasing the efficiency.

II. DESIGN CONCEPT

A. Proposed Process Steps for BC-TWI Solar Cell.
An overview of the process flow is shown in Fig. 2. The via etching is done using a Bosch Etch. The selected diffusion process will use a spin-on glass method. Experiments will be performed to show the effectiveness of the diffusion into the vias.

The outcome of the experiments will determine the type of spin-on glass as well as the diameter of the vias that will be necessary to allow for full diffusion. These results will also finalize the metallization mask design.

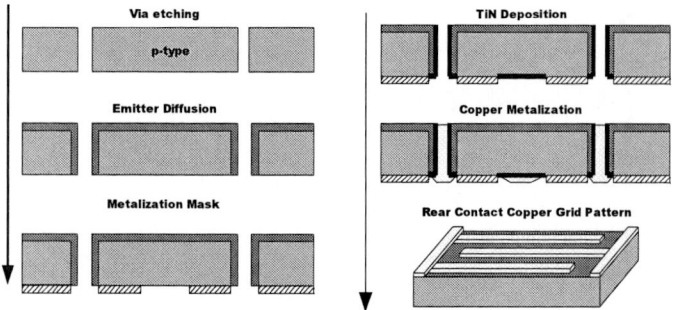

Figure 2 Proposed design and process steps

In other designs, the resistive n-type Si is the primary conductive path through the vias to the rear contact. In this design, the through-wafer interconnects will be the primary conductive path. This will reduce the recombination rate and improve the efficiency over the traditional solar cell. As seen in Fig. 2, TiN will also be selectively deposited to prevent diffusion of the copper into the Si.

B. Further Considerations

An oxide may be used for the metallization mask. By using an oxide, it would eliminate the need to remove the mask and to also deposit an insulation layer. The oxide will provide insulation between the emitter and base contacts in the final grid design.

ACKNOWLEDGMENT

This material is based upon work supported by DARPA through the Space and Naval Warfare Systems Center (SSC) under Award No N66001-05-1-8911. Any opinions, findings and conclusions or recommendations expressed in this publication are those of the authors and do not necessarily reflect the views of DARPA.

CMOS Imager Pixel Design for Space Applications

Mark Elgin, Dede Russell, Matt Katula, Ryan Paulsen, and Dr. Stephen Parke
Department of Electrical and Computer Engineering, Boise State University

Abstract – Our project consists of working with two existing pixel designs for a CMOS imager that will be used in space. The first pixel design consists of a standard three transistor model. The second model consists of seven transistors and incorporates correlated double sampling and the behavior and characteristics of this model will also be examined. The investigation of the behavior of these two designs will be carried out using WinSPICE and Silvaco simulation software. We will enhance the current models by comparing the two simulation software results to improve accuracy. In both the three and seven transistor models we are using double gated MOS transistors which allow for superior amplifier design and a higher quality pixel. A significant portion of the research directed toward efficiently connecting the second gate of each transistor. A unique feature of this pixel is the vertical integration or the stacking of the diodes with through box interconnects between the CMOS and the photodiode.

Characterization of Negative Differential Resistance in Chalcogenide Devices Containing Silver

Armand Bregaj and Kristy A. Campbell
Department of Electrical and Computer Engineering, Boise State University

Abstract — Chalcogenide materials are compounds that contain one of the group VI elements such as sulfur (S), selenium (Se), and tellurium (Te). These materials are presently under research in industry and academia for their potential use in electronic memory devices. An electrical property called *negative differential resistance* (NDR) has been observed in chalcogenide devices containing silver (Ag). NDR describes the I-V characteristics of the device where the resistance has a negative slope. Some of the devices that operate in the NDR region are the tunnel diode, resonant tunnel diode, uni-junction transistor, and Gunn diode. The research work presented in this poster includes the measured I-V characteristics of chalcogenide devices exhibiting the NDR property. The electrical data obtained is compared with existing device models to determine a potential mechanism responsible for the observed NDR in silver-containing chalcogenide devices.

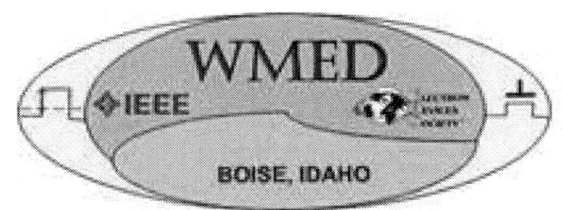

WMED 2006 – Author Index

Ananthan, V. 53
Anser, M. 27

Baker, R. J. 3, 11, 17, 31
Banerjee, P. 55
Belisle, C. 57
Bersuker, G. 31
Black, D. 15
Bregaj, A. 62
Buck, K. 7
Buschick, K. 15
Butt, D. P. 45

Campbell, K. A. 21, 62
Cheek, B. J. 31

DeGregorio, K. 33
Doherty, P. 51
Duvvada, K. 11

Elgin, M. 61
Erbe, A. 59
Estrada, D. 31

Forbes, L. 39, 43

Gonzalez, F. 55
Gorseth, T. L. 31
Gregory, G. 7
Gupta, V. 7

Hackler, D. 33
Harrison, R. 15
Herring, P. K. 21
Hess, H. 51

Hess, H. L. 7
Houle, B. 7
Hwang, D. 29

Iyer, R. 53

Jackson, J. B. 23, 47
Jaeger, R. C. 19
Jessing, J. R. 17, 49
Jun, S.-G. 47

Kapoor, D. 23, 47
Katula, M. 61
Kiepert, J. 31
Kim, S. 15
Klein, M. 15
Knowlton, W. B. 31
Krasinski, C. 35
Krone, S. 51

La Rue, G. S. 9
Louie, M. Y. 43
Luke, M. 45

Marx, B. M. 45
Mcteer, A. 53
Mian, A. 19
Miller, M. S. 23, 47
Moll, A. J. 59
Morinville, W. 35
Mouli, C. 53
Mudrow, M. 39
Naughton, J. J. 27
Noren, K. 51
Normann, R. 7

Ogas, M. L. 31
Owens, T. 29, 53

Palaniappan, N. 53
Parekh, K. 29
Parke, S. 33, 61
Paulsen, R. 61
Plum, T. J. 17, 49
Price, P.M. 31

Ramaswamy, N. 53
Russell, D. 61

Saxena, V. 3, 11, 17, 49
Shoaei, O. 5, 41
Solzbacher, F. 15
Suhling, J. C. 19
Sukumar, V. 51
Surthi, S. 55

Tang, S. 53
Tathireddy, P. 15
Thomason, M. 57
Toepper, M. 15
Tyler, M. 27

Vahidfar, M. B. 5, 41
Vaidyanathan, P. 29

Wanalertlak, W. 39, 43
Wang, S. 53
Warner, D. 51
Williams, B. 57
Wilson, D. 33

Zahller, M. J. 9
Zoschke, K. 15

9781424403738